Elisa MICHEL Delphine SICARD
Bernard ONNO Philippe ROUSSEL

From wheat to bread, the impact of yeast in baking

Elisa MICHEL Delphine SICARD
Bernard ONNO Philippe ROUSSEL

From wheat to bread, the impact of yeast in baking

Effect of baking practices, wheat, flour on biodiversity and yeast activity

ScienciaScripts

Imprint
Any brand names and product names mentioned in this book are subject to trademark, brand or patent protection and are trademarks or registered trademarks of their respective holders. The use of brand names, product names, common names, trade names, product descriptions etc. even without a particular marking in this work is in no way to be construed to mean that such names may be regarded as unrestricted in respect of trademark and brand protection legislation and could thus be used by anyone.

Cover image: www.ingimage.com

This book is a translation from the original published under ISBN 978-620-2-55127-4.

Publisher:
Sciencia Scripts
is a trademark of
International Book Market Service Ltd., member of OmniScriptum Publishing Group
17 Meldrum Street, Beau Bassin 71504, Mauritius
Printed at: see last page
ISBN: 978-620-3-34978-8

Copyright © Elisa MICHEL Delphine SICARD, Bernard ONNO Philippe ROUSSEL
Copyright © 2021 International Book Market Service Ltd., member of OmniScriptum Publishing Group

Sammary

Preamble

This study is part of the various experiments conducted as part of the Bakery contract financed by the French National Research Agency (n°ANR-13-ALID-005), piloted by Delphine Sicard and entitled: "Diversity and functioning of a low-input "Wheat / Man / Microbiome" agro-food ecosystem: towards a better understanding of the sustainability of the bakery sector".

The aim of this Bakery project was to better understand the factors of sustainability of bakery practices in this sector and to disseminate new knowledge for a better control of this sustainability. The implementation of participatory approaches between bakers/researchers was a strong point of the project for a better integration of knowledge.

Major results of the Bakery contract (extract from the final report)

The natural yeasts of French bakers harbour a diversity that has never been described before. The species *Lactobacillus sanfranciscensis* dominates more than 90% of the yeasts. The species *Saccharomyces cerevisiae*, known as baker's yeast, does not dominate the majority of the yeast. Many other yeast species are present, including several of the clade of *Kazachstania*. A new species of yeast, *Kazachstania saulgeensis,* has even been discovered. The genus *Kazachstania* appears as a new model species group for studying domestication. The co-occurrence of bacterial and yeast species does not predict the nature of bacterial-yeast interactions. Depending on the strain, bacteria and yeasts in yeast can cooperate, compete or be neutral for the other.

Two groups of bakery practices stand out: peasant practices and artisanal practices. These practices influence the diversity of yeast species in yeast. Maintaining these two types of bakery practices would therefore help to better conserve biodiversity. The effectiveness of alternative training and knowledge transfer systems contributes to increasing the diversity of baking practices and associated ecosystems.

The environment of the bakery and the baker's know-how mainly contribute to the diversity of the microbial flora of the sourdoughs. There is therefore an effect of the bakery and the baker that could be called a "terroir" effect of the sourdough. The aromatic profiles of the breads change with the soil of the wheat and the soil of the sourdough. The development of a local industry could therefore promote a diversity of tastes. Beyond "health" concerns, consumers are looking for social ties by choosing to eat bread with natural sourdough. This reinforces the interest of local production chains to maintain a sustainable artisanal or peasant bakery.

This synthesis is devoted to the **study of the impact of sourdough on bread quality.** It is intended to be both informative with theoretical reminders (extracts from the Glossary of Knowledge / Breadmaking with natural leaven, placed in separate inserts in the text), elements of the protocols implemented and also descriptive of the results obtained, the observations made and their analysis. To present this work, the choice was made to follow the steps of the approach, associating at each stage the bases of the study protocols implemented with the results obtained and their analysis.

Preliminary remarks :
- Evolution of the <u>Taxonomy of Lactic Acid Bacteria</u>: The names of the lactic acid bacteria mentioned in this article have evolved following a recent study in which a new classification was proposed (Zheng J., Wittouck S., Salvetti E. et *al.,* (2020)). For reasons of simplification, the choice was made to keep here the names of the lactic acid bacteria before this new taxonomic organization.
- The wheat comes from low-input agriculture and this is more directly related to the practices of <u>Organic Agriculture </u>even if the organic label has not been required. The organic bread making is associated with the current practices of the peasant bakers and artisan bakers who work with wheat from an organic agriculture. It also represents bakery practices that rely on fermentation based on the natural biological diversity of the wheat and flour and the environment of the baker without external inputs (additives and various processing aids or selected microorganisms).
- The previous works and bibliography have not been detailed in this synthesis, but can be consulted in the work cited in reference (Michel E. 2018).

Introduction

Bread is the result of a balance between fermentation and dough stability. The properties of the raw material flour depend on various factors (cultivation practices, choice of wheat varieties, milling, etc.) and can therefore be variable. They will influence the characteristics and stability of the dough. The baker, who works with sourdough, has to manage the fermentation activity depending on the microbial biodiversity of the sourdough, but also take into account the biological diversity linked to the enzyme activities of the flour and the microorganisms. The aim of this study is to observe and describe the impact of wheat and sourdough diversity on bread characteristics.

The approach implemented is shown in Figure 1. The presentation of the study then combines the elements of the protocols and the analysis of the results from the main steps:

- Synthesis on the study of microbial diversity of the sourdough panel and choice of experimental bakers;

- Presentation of wheat and flour preparation protocols and their characterization;

- Process of elaboration of experimental yeast by selected bakers and characterization of the yeast;

- Bread-making protocol using experimental yeast and characterization of the doughs and breads obtained.

Levain chef	**All-point sourdough**
Portion of dough taken from the sourdough or the day's kneading which serves as a starting base for the preparation of the sourdough any point used as a fermenting agent. The remaining part of the leaven is preserved, its activity is maintained in a regular way by refreshment (supply of	Sourdough prepared to ensure the fermentation of a kneaded dough. Through a succession of refreshments, the quantity of sourdough from a master sourdough is gradually increased while maintaining its microbial activity in optimal phase.

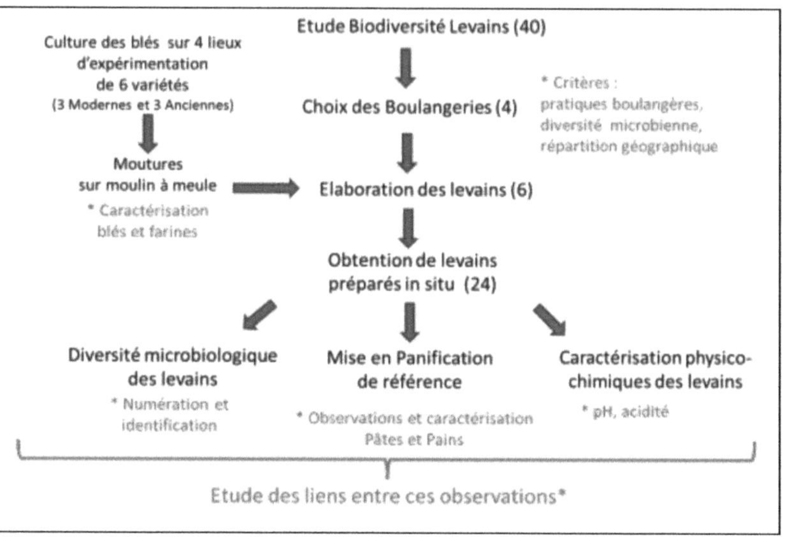

Figure 1: Bakery experimental device - Sourdoughs, Wheat and Soils

In a first step, a characterization of the microbial biodiversity of sourdoughs on French territory was carried out on 40 different sites at farmers (16) and artisan bakers (24) who exclusively practice sourdough bread making with wheat from Organic Farming. The aim was to gain a better understanding of the influence of bakery and agricultural practices in the implementation and monitoring of sourdough in bread-making.

In a second step and within this panel of bakers, a selection of 4 types of sourdough, differentiated by the microbial diversity of the yeast and bacteria populations, was selected. The 4 bakers chosen on the specificity of their sourdough volunteered to initiate 6 master sourdoughs (L1 to L6), in their working environment and with their practices, from 6 flours from a mixture of modern varieties and a mixture of varieties from ancient wheat populations, each cultivated in 3 different terroirs. The aim was to observe the effect of these factors (baker, flour, terroir) on the characteristics of the leavens elaborated.

In a third step, from these starter cultures, all-point sourdoughs (see inserts) were prepared for the implementation of breadmaking trials with a reference diagram using the flours from the study. The aim was to characterize the impact of these sourdoughs during bread-making and on the breads by observing the properties of the dough and by measuring acidity, gas production and the dosage of chemical components on the breads.

1. Characterization of the microbial diversity of the yeasts of peasants and artisanal bakers in the bakers' panel

1.1. Protocol for characterizing the microbial diversity of sourdough in the baker panel

A **leaven** is a microbial ecosystem made up of lactic acid bacteria and yeasts interacting with each other and with the cereal environment. This medium is itself the seat of endogenous enzymatic activities (proteases, amylases...) influenced by microbial activities. This ecosystem is said to be complex because of the diversity of the microbial flora present and the interactions with the flour medium, which can vary in terms of composition (Figure 2). It is generally accepted that a leaven can contain 1 billion lactic acid bacteria and 10 million yeasts per gram. A sourdough is therefore to be considered as an addition of fermentative activity to the dough, but also as a pre-fermentation stage for a fraction of the dough, with modification of the constituents and their properties (addition of acidity, hydrolysis or solubilization of proteins, release of flavour precursors, etc.).

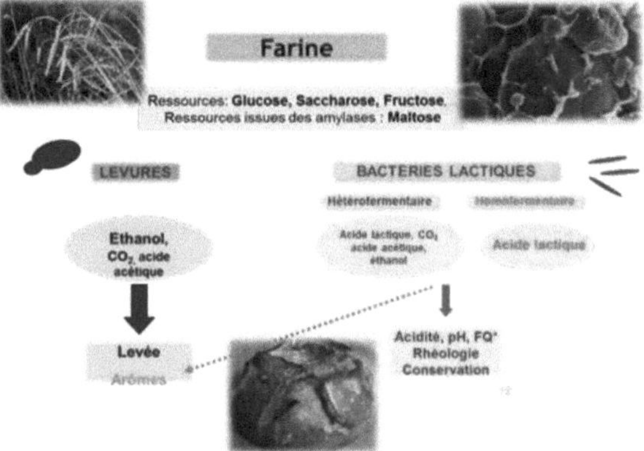

Figure 2: Sourdoughs, complex cereal ecosystems

The Recueil des Usages des Pains en France, (CNERNA, 1977), states for sourdough: "Dough in acid reaction fermentation, initially coming from a mixture of flour and water, without the voluntary addition of yeast and perpetuated from this mixture, once it has undergone spontaneous fermentation, by additions conducted in a methodical way". More recent work and writings (Gobbetti and Gänzle, 2013) have led to the notion of a complex microbial ecosystem to describe yeast (see Box and Figure 2).

The characterization of the microbial diversity of the yeast from natural fermentation and of the metabolites formed during fermentation is based on a protocol, summarized in Figure 3, for the genetic identification of the yeast microflora from cultural and non-cultural methods and from physico-chemical analyses of yeast and bread.

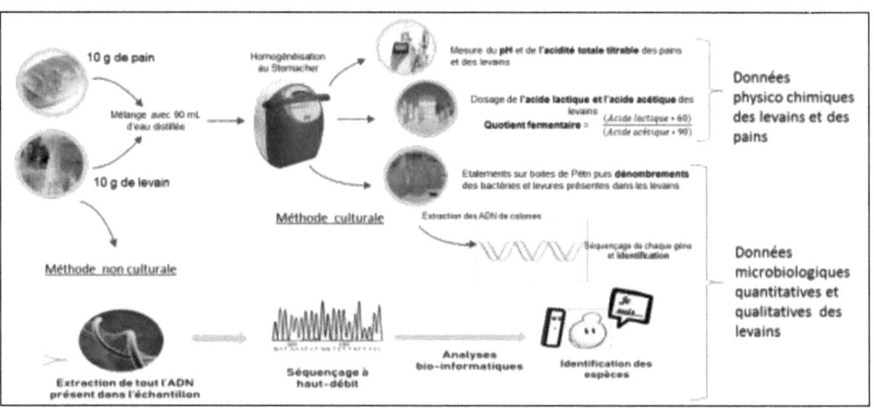

Figure 3: Methodological process of physico-chemical and microbiological analyses of yeast and physico-chemical analyses of breads

1.2 Microbial diversity and density of sourdough in the bakery panel

The results concerning the density of microbial populations in the different yeasts in the baker panel are presented in Table I.

Table I : Enumeration of bacteria and yeasts in the bakers' panel (40 yeasts) (UFC : Unité Formant Colonies)

Microorganisms	Average microbial density	Differences observed
Yeast	$2.16. \ ^{107}UFC/g$	$3,58. \ ^{105}$ to $9.23. \ ^{107}$ CFU/g
Lactic Bacteria	$1.23. \ ^{109}$ CFU/g	$3,60. \ ^{105}$ à $4,17. \ ^{109}$ CFU/g

These quantitative results are in agreement with the data of many authors on average densities of yeasts and lactic acid bacteria (Onno and Roussel, 1994; Gobbetti and Gänzle, 2013). They also show the variations in microbial density from one sourdough to another: variation factors from 1 to 300 for yeasts and from 1 to 10,000 for lactic bacteria (Table 1). The ratio N yeasts / N lactic acid bacteria is between 1/100 and 66/100. The origin of these quantitative variations is related to the conduct of the yeasts (T°C, frequency of refreshments ...) and also to the physiology of microorganisms present, some of which are non-cultivatable in enumeration media.

Like
The species are grouped into genera. Species that belong to the same genus are genetically closer than those that belong to different genera. The genus is given by the first name of a species. For example the species *Saccharomyces cerevisiae* is in the genus *Saccharomyces*. The soft wheat species, *Triticum aestivum,* is of the genus *Triticum.*
Species
In micro-organisms, a species is a group of individuals linked by close family ties: they therefore have a genome (DNA) and very similar morphological and physiological characteristics. In bacteria there is no sexual reproduction. In yeasts, it depends on the species. When there is sexual reproduction, individuals of the same species are able to reproduce among themselves and give fertile offspring.
Strain
In microorganisms, a strain is in some ways the equivalent of a "homogeneous variety" for cultivated plants.

The diversity of lactic acid bacteria species of these yeasts has been studied (Lhomme et al, 2016, Michel et al, 2016). Among the 19 species of lactic acid bacteria identified in all yeasts, *Lactobacillus sanfranciscensis* is confirmed as the predominant species. For sourdoughs from baker farmers, 5 species were found, including 2 species specific to these sourdoughs (L. *brevis, L. paralimentarius*). As regards artisan bakers, 17 species were found, including 14 species specific to these yeasts. By yeast, 1 to 5 species of lactic acid bacteria were identified. There does not appear to be a link between the diversity and geographical distribution of sourdough.

These main results are consistent with the data in the literature. However, the differences observed here between the sourdough of peasant bakers and artisanal bakers are worth mentioning.

For the yeasts, 766 strains (see insert above) were identified and 40 yeasts were analyzed by metabarcoding (see insert below). The number of species is 8 for farmer bakers and 10 for artisanal organic sourdough bakers. By sourdough, 1 to 3 dominant species are identified. Three majority species are present: *Saccharomyces cerevisiae, Kazachstania humilis, Kazachstania bulderi.*

An effect of baking practices is observed. Indeed, *K. bulderi* is common among peasant bakers and *K. humilis is* common among artisan bakers. The genus *Kazachstania* constitutes a new model for studying the evolution of yeasts and reflecting on the conservation of biodiversity. It is worth noting the identification of a new species *Kazachstania saulgeensis*, and the presence of two species recently discovered in other environments

Metagenics

Meta-genetics is an independent culture method for characterizing the diversity of microbial **species** in an **ecosystem**. It is said to be independent culture because it does not require the isolation of individual strains of **bacteria** or **yeasts** from the **yeast.** The DNA is extracted directly from the yeast and therefore from all the microbial species of the yeast at the same time. A gene or a region of a gene is then amplified and sequenced from this DNA. This results in a large number of sequences that come from the different microbial species of the **yeast**. These sequences are assigned to species on databases, which makes it possible to know the species composition of the sourdough.

1.3 Characteristics of the sourdough and breads in the bakers' panel

The results obtained on the acidity characteristics of the breads in the panel show strong variations in pH values and acetic acid content (Table II).

Table II: Sourdough pH, pH and acetic acid content of breads from the bakers' panel

	pH yeasts	pH breads	Acetic acid in breads (ppm or mg/kg)
Average value	3,80	4,48	740
Differences observed	3.54 and 4.03	3.91 and 4.85	420 and 1200

The values obtained do not always respect the recommendations of the 1993 bread decree. Indeed, this decree, in its modified version for sourdough bread, specifies that traditional French sourdough bread must have a pH < 4.3 and an acetic acid content > 900 ppm (mg/kg). Here again, these results point out the limits of this regulatory text.

1.4. Selection of experimental bakers for the study of the impact of sourdough in baking

The analysis of microbial diversity and bakery practices allowed us to select 4 organic sourdough bakers (B1, B2, B3, B4), which are very distinct by the dominant flora of the microorganisms present in their own sourdoughs (Table III), and also by their type, activity and terroir: a farmer (B1, department 86), an artisan (B2, department 44), a peasant (B3, department 28), a craftswoman (B4, department 68). None of these bakers use baker's yeast.

Table III: Microbiological characteristics of homemade or baker's yeast (B1, B2, B3, B4) obtained by classical microbiology from cultured and isolated strains of sourdough (Urien et al. 2019; Michel et al. 2016; Michel et al. 2018)

Bakers	Main species of yeast microorganisms	
	Main yeast species (K.:Kazachstania, Saccharomyces) S.:	Main species of lactic acid bacteria (L: Lactobacillus) In bold: heterofermentative species required In non-bold: optional heterofermentation
B1	K. saulgeensis (97.5%), S. cerevisiae (2.5%)	**L. sanfranciscensis** (100%)
B2	K. bulderi (100%)	L. sakei (46%), L. **sanfranciscensis** (54%)
B3	S. cerevisiae (96%), K. unispora (4%)	L. curvatus (90%), L. plantarum (10%)
B4	S. cerevisiae (91%), Kazachstania sp. (9%)	**L. brevis** (69%), L. plantarum (15%), **L. sanfranciscensis** (8%), L. paralimentarius (8%)

2. Protocols for the preparation and characterization of raw materials for bread-making tests

2.1. Wheat cultivation protocol

The wheat was sown in autumn 2014, harvested in July 2015 at 4 farmers (places of experimentation). We distinguish the 3 wheat populations - Redon population, Saint Priest population and Bladette population cultivated on 3 terroirs - and the 3 modern varieties (Chevalier, Renan, Pirénéo,) cultivated on 3 terroirs. The populations of ancient wheat from the same terroir are mixed in equal parts to form an Ancient Mix associated with one terroir, the operation is repeated for the other two terroirs. The same process is carried out for modern varieties in the constitution of the Modern Blends.

The description and coding of raw materials are detailed in Table IV.

Table IV: Identification of wheat and flour intended for breadmaking trials (MA Ancient Mix, MM Modern Mix)

Cultivation		identification		
Places of experimentation	Classification	Varietal mixtures	Flour (Astrié version)	Levain Chef
GS (35)	Old Mix	GSMA	F1	L1
LA (35)	Modern Blend	AML	F2	L2
FM (49)	Old Mix	FMMA	F3	L3
GS (35)	Modern Blend	GSMM	F4	L4
LA (35)	Old Mix	LAMA	F5	L5
LM (53)	Modern Blend	LMMM	F6	L6

2.2. Wheat milling protocol

The sorting of wheat after harvest was carried out either at the farmer's with a cleaner-separator or at INRA (air column) supplemented by an optical sorting at a farmer's premises.

The wheat before milling was stored either at the farmer's, in a shed, or at INRA, in a dry room at 20-25°C until the end of October and then at 10°C.

For wheat milling, the Ancient Mixes (MA) and the Modern Mixes (MM) were delivered on the same site, to one of the selected bakers. The milling, in a single pass, on a 50 cm diameter Astrié type grinding wheel (Figure 4), followed one another, alternating the Ancient and Modern Mixtures; the wheat was not moistened before milling. The sieving was carried out in a sieve with a mesh opening of 300 μm; after each milling, the sieve was cleaned by suction. The batches of flour were then bagged (Figure 5).

Figure 4: Astrié Mill	**Figure 5**: Flour Sampling	**Figure 6**: Palletization of samples

Sample preparation was organized to ensure the regularity and traceability of sampling throughout the process by each baker. Each pallet (Figure 6), intended for the 4 bakers, contains :

- 12.5 kg for each of the 6 flours, in bags of 5 and 2.5 kg, numbered from 1 to 6;

- 6 buckets with lids to initiate and maintain the 6 yeasts for 3 weeks ;

- 6 small buckets with lids and labels to put 1 kg of chef ;

- Sterile vials for all samples.

3. Characterization of wheat and flour

3.1 Characterization of wheat hardness level

Wheat hardness

All the wheat used is soft wheat of the *Triticum aestivum* species. Among the soft wheat varieties, the wheat can be separated according to the hardness of its grain. The objective of this qualitative characterization was to clearly identify the "Soft" and "Hard" characteristics of the Ancient and Modern Blends that appeared with the genetic evolution of wheat. The hardness in the scale of values of the official method is based on a flour extraction rate after milling on a selected mill (KT mill). A more qualitative picture of the particle size distribution is obtained by means of the grain size profile and the average size of the flour grains.

From a technological point of view, the analysis of the hardness level allows to answer the question of preparation of wheat for milling. The "hard" character of wheat justifies humidification of the wheat in order to allow a good separation of the grain envelopes from the starchy albumen and to limit the friability of the envelopes to avoid a qualitative variation of the flours. In fact, crumbly husks lead to a greater quantity of fragments of this histological part of the grain in the flour. The husks are better dissociated from the albumen and give larger sounds. The presence of these husks in the flour is determined by the ash content.

The hardness measurement method used was based on the study by Triptolème, (Triptolème, 2017-1) which validated a simple milling protocol to obtain flour particle size profiles similar to those obtained with Astrié millstones and Chopin laboratory rolls. In our study (Table V), the sieve size characterization was carried out in "mastersizer 2000" laser sieve size. Wheat grinding was carried out on a manual coffee grinder of 25 g ± 0.1 g. The extraction rate of the flour extracted on a 180 µm sieve is, on average, 19.5% for the Modern Mixes and 23.5% for the Ancient Mixes.

Table V: Grain size characteristics of wheat flours from the grain hardness test (laser grain size) (MM: Modern Mix, MA Old Mix in gray)

Codes varietal mixtures	Moisture content (%)	Average flour particle diameter (µm)	d (0.1) *	d (0.5) *	d (0.9) *
F1 (GSMA)	12,92	56,1	7,9	35,2	134,3
F2 (AML)	13,57	68,0	11,7	53,3	146,9
F3 (FMMA)	14,84	54,6	8,0	34,2	130,7
F4 (GSMM)	13,24	60,6	10,2	42,6	137,6
F5 (LAMA)	12,93	58,0	8,2	37,1	138,0
F6 (LMMM)	15,15	66,5	12,0	49,9	145,9
Average MA	13,6 ±0,9	56,2 ±1,4	8,03 ±0,1	35,5 ±1,2	134,3 ±3,0
Average MM	14,0 ±0,8	65,0 ±3,2	11,30 ±0,8	48,6 ±4,5	143,5 ±4,2

* diameter (d) below which are located 10%, 50% and 90% of the particles characterized by their volume

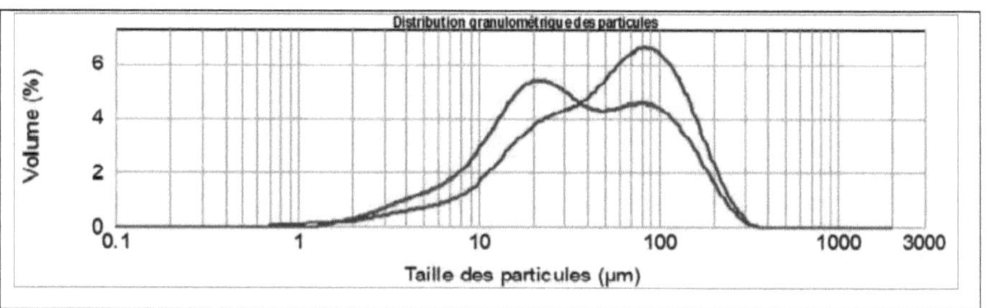

Figure 7: Particle size profile of a mixture of old varieties (LAMM in red) and a mixture of modern varieties (LAMA in green)

The higher average particle size (Table V and Figure 7) and the monomodal tendency profile of the particle size curves for Modern Blends (green curve) compared to the Older Blends, with a finer average particle size and the bimodal tendency profile of the curves (red curve), confirm the more pronounced hard character of the Modern Blends. The red curve can be described as a soft medium and the green curve as a hard medium, considering a hardness scale between "very soft" and "very hard" (Triptolème, 2017-1).

In milling technology, as the level of hardness increases, the ability of the grain to break down into flour is reduced and the grain size is increased. This characteristic of the grain is partially corrected by moistening it to facilitate its reduction. For this study, the moderate differences in the level of hardness between the Modern and Ancient Mixes and the higher average moisture content of the Modern Mixes (Table V), led us not to moisten the grains for milling wheat for breadmaking.

3.2 Characterization of Astrié Mill Flour

Table VI: Characteristics of flours for baking trials
(MM: Modern Mix, MA Ancient Mix in gray)

Flour Codes for varietal mixtures	Extraction rate (%)	Content in water (*) (%)	Content in ashes (**) (%)	Rate of damaged starches (***) (UCD)****
F1 (GSMA)	83- 85	12,8	0,69	20,2
F2 (AML)	83- 85	12,7	0,69	24,85
F3 (FMMA)	83- 85	12,8	0,8	21,8
F4 (GSMM)	83- 85	13,0	0,8	22,25
F5 (LAMA)	83- 85	13,0	1,0	21,33
F6 (LMMM)	83- 85	13,1	0,7	25,4
MA (F1, F3, F5)	-	12,9 ±0,1	0,83 ±0,13	21,1 ±0,7
MM (F2, F4, F6)	-	12,9 ±0,2	0,73 ±0,05	24,2 ±1,4

(*) Standard NF V03 707, (**) Standard NF V03 720, (***) Standard NF V03 731, (****) Chopin Dubois unit

The yields or extraction rates obtained for type 80 are in the same range as those obtained by Triptolème (2017-2) in the PaysBlé research contract with this type of mill.

Ash contents are between types 65 and 110 (Table VI) and randomly between Old and Modern Blends. The means and standard deviations therefore do not distinguish between these mixtures; each value is the result of a mean of two 2 g samples and shows good repeatability. Considering the reliability of the Astrié milling and the influence of hardness on this factor which gives often higher ash contents for modern varieties compared to old population varieties, there remains an uncertainty in the sampling. Indeed this measurement is very much impacted by effects of particle classification in a lot. The particles from the periphery of the grain, which are less dense, separate quite easily from the albumen particles and are positioned rather in the upper part of the packaging which would have justified the implementation of a protocol re-homogenization of the lot before analysis. The rate of damaged starches is higher with Modern Blends compared to Ancient Blends (Table VI). These results are consistent with the relationships established between this measure and wheat albumen hardness. Indeed, the harder the grain is, the more resistant it is to passing between the two millstones and the less easily it fragments; areas of breakage occur more randomly even in starch grains.

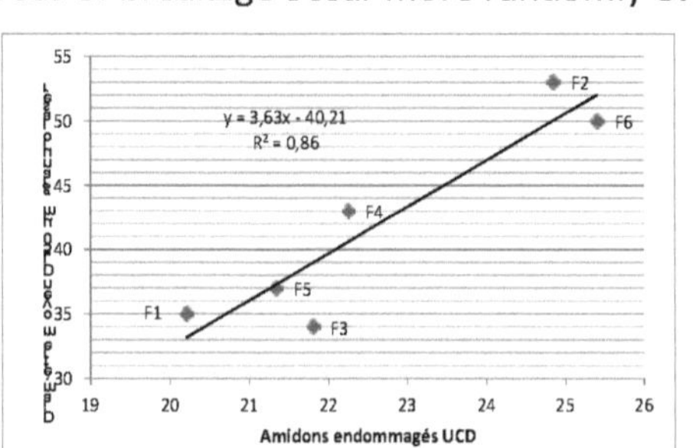

Figure 8: Relationship between flour particle size (analysis on wheat) and damaged starch content of flours (Astrié milling)

Consequently, there is a relationship (Figure 8) between damaged starch particle size and variety categories (Modern Blends and Older Blends), explained in Chapter 3.1. These results are consistent with the previous results from the WheatCountry contract (Triptolème 2017-2).

3.3 Characterization of gluten

The term "**gluten**" comes from the Latin "glutinum" which means bond or glue and only wheat proteins (gliadins and glutenins) have the property of strongly associating or sticking together in the presence of water. Gluten does not exist as such in flour, it is formed in the presence of water in the dough. Wheat gluten thus ensures the formation of a continuous protein structure which is capable of retaining fermentation gases that allow the dough to rise in the making of bread. This property gives the wheat the qualification of bread-making quality.

Wheat gluten is a complex, elastic and variable association of proteins of high molecular weight in a hydrated medium, formed by :
- polar or hydrophilic (with water) connections ;
- hydrophobic interactions (with fats) ;
- Ionic bonds, some atoms being electrically charged + and - (bonds with charged elements such as salt or acids ...) ;
- oxidation bonds between cysteine molecules (high-energy bonds).

The diversity of protein types and the bonds between these proteins contribute to the different quality characteristics of baker's doughs. The elastic strength and extensibility of gluten is reduced by reducing compounds and hydrolysis due to the activities of the flour proteases, which occur to a very significant extent in baking yeast. It definitively loses its deformability during thermal coagulation during baking.

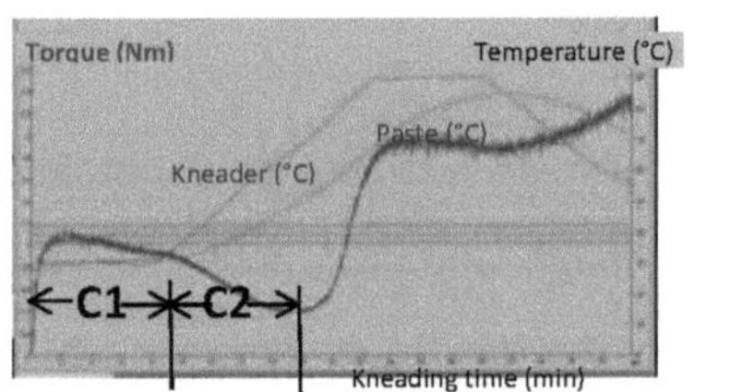

| Figure 9: Manual leaching for the extraction of Moist Gluten (GH) | Figure 10: Centrifuge (Glutomatic) with sieve for the determination of the Gluten Index (GI) | Figure 11: Curve (green) of resistance (Nm) of a Mixolab paste as a function of temperature (C2 = average resistance between 30°C and 60°C) |

The extraction of the moist gluten (GH) from the dough was done manually (Figure 9) using the ITCF method (1972), which is better suited for high-fibre flours than the Glutomatic apparatus (Figure 10). With a water content of 50% of flour, the hand-formed doughs with Modern Mixes required more energy and more work for agglomeration due to their stronger water binding capacity. The results show a higher content (proportion expressed in relation to the flour) of wet gluten in the Old Mixes compared to the Modern Mixes (Table VII). Under the experimental conditions applied, the Ancient Mixes lead to rapid aggregation and therefore to the obtaining of a homogeneous, well-bound gluten; with modern wheat, a heterogeneous and not very extensible gluten mass is observed.

The Gluten Index (GI) measurement at the Glucomatic (Figure 10) integrates the capabilities of wet gluten in forced passage, by centrifugation through a sieve. When the passage rate is high, the index tends towards the value 0; conversely for an impossible passage, the index takes the value 100, it is related to a high level of aggregation, a sign of high elastic resistance. The qualifiers "soft", "less aggregative" are often associated with low index glutens. Gluten Index (GI) values are lower for Older Blends (Table VII). It has also been noted that during centrifugation a higher water loss was observed with the Ancient Mixes compared to the Modern Mixes.

The Mixolab was chosen to obtain additional information on the resistance of the dough and thus indirectly on the gluten during kneading at different temperatures. Dough formation is carried out at a temperature of 30 °C and for a constant resistance torque C1 by adjusting the hydration. The value of torque C2 (Fig. 11) was chosen because it reflects a weakening of the gluten network after kneading under the effect of temperature (between 30 and 60 °C) and before the phenomenon of dough thickening by gelatinization from 60 °C onwards. The C2 torque is lower with the Ancient Mixes (Table VII), reflecting a lower gluten resistance in the temperature range of 30 to 60 °C.

The fractionation of proteins and their separation in chromatographic method (HPLC) allows to show for glutenins (fractions with very high molecular weight) a higher proportion than gliadins with Modern Blends: this results in a higher glutenin/gliadin ratio. These results confirm the observation of professionals that the elastic resistance of pasta with modern wheats is greater than that of old varieties.

The set of physico-chemical data on mixtures of ancient and modern population varieties allows to distinguish them well by the analytical methods used. The results obtained give for the Ancient Mixtures, compared to the Modern Mixtures :

- a lower albumen hardness of the grain;
- a lower rate of damaged starches ;
- a finer flour grain size ;
- a higher average protein level and a higher gluten level ;
- a lower level of gluten and flour dough strength.

These observations are in line with those recorded in the PaysBlé contract (Tripto, 2017-2).

Table VII: Protein characteristics of wheat and flour for breadmaking trials. (MM: Modern Mix, MA Old Mix in gray)

Varietal Mix (*)	Protein on wheat on wheat % on dry matter	Hagberg on wheat (**) Time from fall(s)	Flours (***)	Chromatography in HPLC % Gliadines	% Glutenins	Glut/Glia ratio	Gluten dosage Wet Gluten (%) ****	Gluten Index (%) *****	Dough resistance to kneading couple C2 (N.m) ******
GSMA	11,3	386	F 1	43,2	38,8	0,9	26,0	44	0,32
AML	10,2	363	F 2	40,3	41,3	1,02	20,0	89	0,48
FMMA	12	402	F 3				29,3	39	0,36
GSMM	10,7	385	F 4	40,1	41	1,02	18,7	87	0,38
LAMA	12	412	F 5	43,6	38,9	0,89	28,4	62	0,31
LMMM	12	405	F 6				25,5	64	0,51
Average MA	11,77 ±0,3	400,00 ±10,7		43,40 ±0,2	38,85 ±0,05	0,90 ±0,01	27,90 ±1,4	48,33 ±9,9	0,33 ±0,02
Average MM	10,97 ±0,75	384,33 ±17,15		40,20 ±0,1	41,15 ±0,15	1,02 ±0,0	21,40 ±2,95	80,00 ±11,3	0,46 ±0,06

(*) GS, LA, LM: terroir identification; MA, MM: mix of old and modern varieties

(**) Standard NF V03 703 viscosity index related to the amylasic activity of wheat (the higher the value in seconds and the lower the activity, the reference values are between 250 and 300 s)

(***) Flour extracted from the Astrié milling varietal mixes

(****) Manual extraction, ITCF protocol, 1972, (Figure 9)

(*****)ISO Standard 7495, Glutomatic Perten, (Figure 10)

(******) Standard NF V03 765, Mixolab Chopin, (Figure 11)

4. Preparation and characterization of experimental starter cultures for bread-making trials

The mission of the selected bakers (B1, B2, B3, B4) was to initiate and maintain sourdough according to the specific practices of each baker by respecting the recommendations, mentioned in the follow-up sheet, to avoid contamination between sourdoughs (containers and clean hands), but identical for the 6 flours (F1 to F6). Each baker found on his pallet (Figure 6), 12.5 kg of each flour, buckets to initiate, maintain and refresh the sourdough, "sourdough pots" for repatriation, a sheet for monitoring the sourdough (dates, recipe, etc.) and sampling tubes for analytical monitoring (Figure 3).

The start of the sourdough over a period of 3 weeks was carried out with different parameters depending on the bakers. At the same time, a control series of sourdoughs (B5) was developed under laboratory conditions. The main factors are summarized in Table VIII. The number of refreshments ranged from 5 to 13. The time between each refreshment ranged from less than two per week to at least four per week. The average storage temperatures during the sourdough-making period also varied from baker to baker. The average quantity of flour used differed from baker to baker, as did the hydration of the sourdough between 81 and 100% on a flour basis. Finally, the quantity of sourdough used at each refreshment was either constant over time (e.g. 50% for B1 and B2) or variable (B2 and B3). This parameter has an impact in particular on the average acidity level of the sourdough over time; a high level of sourdough leads a priori to more acidic conditions. It should be noted that no baker's yeast was used in these 4 bakeries.

Table VIII: Elaboration of sourdoughs, synthesis of refreshment practices by baker

	B1	B2	B3	B4	B5*
Total number of refreshes	5	7	9	10	13
Average number of refreshments / week	<2	<3	~3	>3	>4
Start-up hydration (% on flour basis)**	100	86	100	88	100

Average flour (g) by refreshed	433	200	646	566	200
Medium Hydration (% flour basis)	100	81	81	85	100
Average fermentation temperature (°C)	14°C	23°C	12h 24°C then 48h 8°C	21°C	20°C
Quantity of sourdough during refreshment (% / total dough)***.	50%	50 à 35%	20% then 10%.	50%	30%

*B5: Sourdough control series developed in the laboratory
**% on a flour basis: quantity of water used in relation to the flour on a 100 basis
*** Either constant quantity or change during refreshes

The sampling protocol for the analyses was organized over the 3 weeks defined for the elaboration of the starter cultures (J0, s_1, S2, S3). On the day of initiation of the sourdough (D0) and each weekend (s_1, S2, S3), two samples per sourdough were taken and all were frozen except for S3. In addition, 2 samples of homemade sourdough (LM) were taken at the beginning and end of the sourdough development to evaluate their potential variations during the experiment. Frozen or fresh samples were sent to the analysis laboratories under isothermal conditions on the same day, at the end of S3, for the 4 bakers. A sample of the S3 yeast was directly analyzed for its yeast and bacteria content. A sample of each sourdough was stored at -80°C for subsequent microbiological and biochemical analyses. The S3 sourdoughs were used for the experimental bread-making process.

4.1. Monitoring of microbial flora at the different stages of development of experimental yeasts

Meta-genetic analysis (see Box §1.2) focuses on the implantation of the different genera and species of bacteria and yeasts during the elaboration of yeast, from grain to yeast. This analysis is carried out by a non-cultural method using DNA extracted from samples and high-throughput sequencing of genes specific to bacteria on the one hand and to the fungi to which yeast belongs on the other. The relative abundance results for the identified genera are shown in Figure 12.

At the grain stage, there seems to be little difference between the 36 samples analyzed, neither between wheat varieties nor between terroirs (data not shown). At the flour stage, 24 batches were prepared from wheat mixtures and again, the same genera in terms of bacteria and fungi were found in these samples, without any effect on varieties or terroirs.

At the yeast stage, the monitoring took place from the elaboration J0 and during three weeks (S1, S2, S3). Progressively, majority flowers appeared.

For bacteria, it is the genus Lactobacillus (dark blue color). This genus becomes progressively dominant in all S3 yeasts. The genera (Pseudomonas, Bacillus, in particular, red and pink color) present in the grains and flour disappear in the course of time. The genus Lactobacillus is present but in very small proportions in flours.

Note that the 6 samples of B1 (left at each stage) show a different bacterial flora from the other samples (at stages S2 and S3 in particular), with more bacterial genera present at 3 weeks, probably due to the low number of refreshes (5) for this baker.

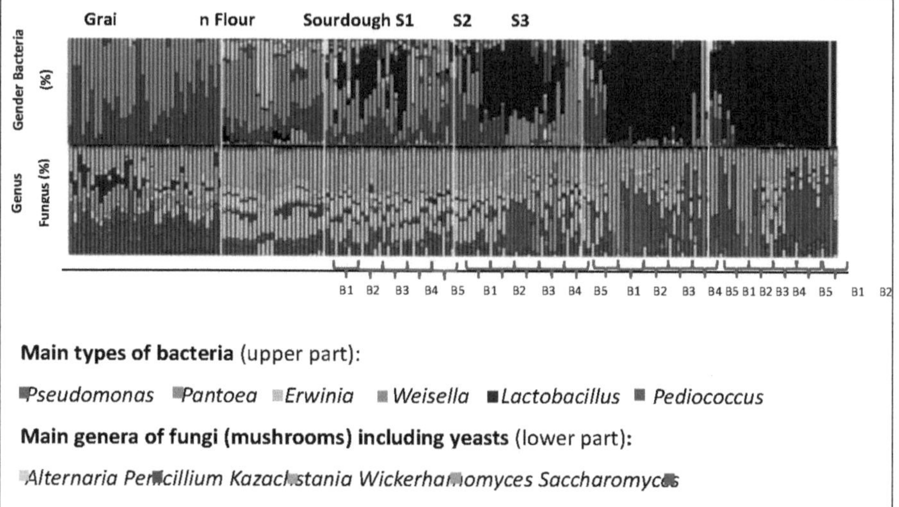

Figure 12: Relative abundance (%) by non-crop method, of the different genera of Bacteria and Fungi including Yeasts during the different stages of yeast development (Each bar/column represents a sample of wheat, flours and yeast: Grains N=36, Flours N=24, Sourdoughs from D0 to Week S3 made from 6 flours, in the order of bakers B1, B2, B3, B4, and B5 for laboratory testing).

For yeasts, the evolution seems slower, with the gradual appearance in S1 of the genera *Kazachstania* (light green color) and *Saccharomyces* (red color), *Wickerhamomyces* (salmon color). Compared to bacteria, a greater diversity remains in S3.

The DNA of fungal flowers corresponding to filamentous fungi (*Alternaria* for example ochre color) present in the grains seems to be detected at this stage.

This approach, by direct analysis of the microbial DNA of the sourdough, gives a global picture of the implantation and selection dynamics of the majority of microbial flora. It is completed in paragraph 4.22, by a cultural approach allowing both to quantify the two populations, yeasts and bacteria, and to identify the species present within each group.

4.2 Characterization of experimental yeasts

After the last refreshment, S3 at the bakers, and after storage for 24 hours at 4°C, the enumeration and identification of the yeast and bacterial flora of the different yeasts were carried out. The determination of the total titratable acidity and the pH measurement were also carried out at this stage.

4.2.1. Enumeration of microbial flora in experimental yeasts

The results of the lactic acid bacteria and yeast counts of the different yeasts are shown in Figure 13.

The range of variation for the number of lactic acid bacteria in experimental yeast is between 2.2.108 and 2.9.109 CFU/g, a factor of 10. The density of lactic acid bacteria depends on the baker's practices. It should be noted that the values obtained in the laboratory with 13 refreshes reach an average of 6.8.109 CFU/g. Sourdoughs B3 have values 4 times lower on average than the other sourdoughs B1, B2, B4, the cold storage of the sourdoughs for B3 between two refreshments may explain this observation. With the exception of B1, the average lactic acid bacteria count values of the experimental sourdoughs are close to those of the homemade sourdough from the same baker.

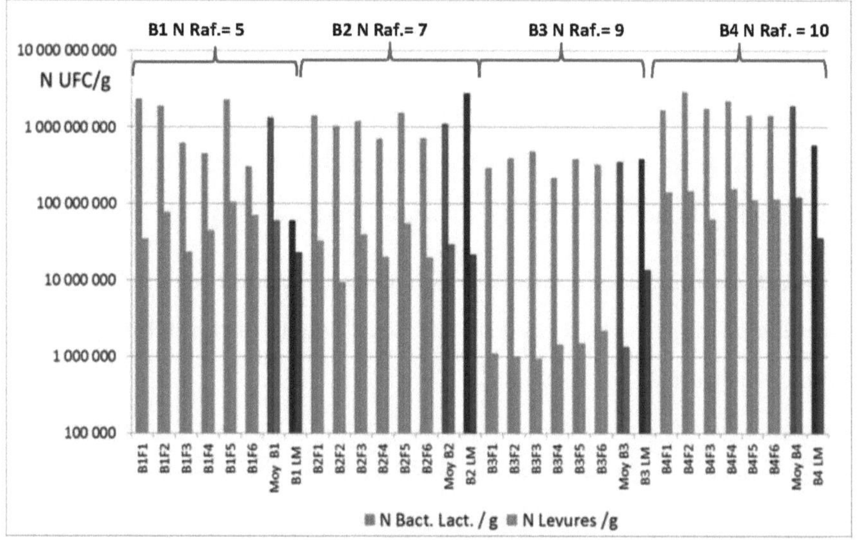

Figure 13: Density of yeast and lactic acid bacteria populations in the different sourdoughs (B1 to B4) with the different flours (F1 to F6) and homemade sourdough (LM) or used by each baker. (N CFU / g : Number expressed in Colony Forming Unit / g) (N Raf. : Number of refreshments / baker at the top of the graph)

- Average N Lactic Bacteria / Baker's Bacteria
- Average N Yeast per baker
- N Lactic Bacteria Homemade Sourdough
- N Homemade Sourdough Yeast

For **yeasts**, the microbial density varies from $1.^{106}$ to $1.6.^{108}$ CFU/g, a factor of 100. The variability is therefore higher than for lactic acid bacteria. Comparing these results with those of Figure 13, it is possible to hypothesize that the yeast microflora is not yet qualitatively and quantitatively stabilized at this stage (S3). As for lactic acid bacteria, yeasts are significantly lower in number for B3, (possible link with cold storage). Compared to homemade sourdough, B3 sourdough has an average value 10 times lower. The highest microbial density levels are reached for B4 ($1.^{108}$ CFU/g), with a number of 10 refreshed. The values obtained for laboratory-produced sourdough, with 13 refreshes, are close on average to that of B4 ($8.6.^{107}$ CFU/g). However, the number of refreshes does not seem to explain the variations in density alone. In fact, with few refreshes, B1 yeasts show higher values ($6.^{107}$ CFU/g) than B2 and B3.

Looking at both microbial populations, B3 yeasts show significantly lower bacterial and yeast populations than yeasts from other bakeries, and for B4 yeasts a tendency to higher values.

A baker's effect is therefore well observed on the microbial density of the sourdough. The higher microbial density for B4 yeast, and also B5 yeast, may be the result of a particular species composition or a higher number of refreshments that lead to a sourdough pH, nutrient supply and oxygenation level more favourable for microbial multiplication. The low microbial densities of B3 sourdough can be attributed to storage at 4°C between each refresh. The lower temperature and the lower number of refreshments for B1 do not seem to have penalized microbial development.

4.2.2. Nature of the microorganisms present in experimental yeast

The majority of the yeast flours are shown in Table IX, without distinction of the flours used, and in Figure 14, taking into account the 6 carrier flours. The microflora of the home-made sourdoughs is given for the purpose of comparison with the sourdoughs used in the experiment.

For homemade sourdough, there are some differences from the initial analysis of sourdough during the baker selection phase (Table III), showing that the sourdough flora can change over time (Table IX).

For the experimental sourdoughs, the flours of the chef sourdoughs initiated by the bakers are quite close to the homemade sourdoughs (LM) or the usual sourdoughs of these bakers. The term "House microbiota" is used to describe this flora present in the bakery and linked to the practices of the baker, which could therefore have become established in the new sourdoughs prepared on site. It seems that the more refreshments there are (from 5 to 10), the closer the flora of the sourdough prepared for the experiment is to that of the home-made sourdough (Figure 14). However, the microflora of the Labo yeast (B5), initiated in the laboratory in an aseptic environment, remains fairly diversified despite the large number of refreshes (13).

As for the yeast species present, *S. cerevisiae* is the majority species for B1 and B4, as well as for the B5 yeast prepared in the laboratory, while as previously mentioned, there is no use of baker's yeast in the 4 bakeries. However, since B1 and B4 are opposite in terms of the number of refreshments (5 to 10), the predominance of *S. cerevisiae* in the sourdough seems to be more related to the "household microbiota" than to the conduct of the sourdough. These results confirm the observations made by non-cultural methods (Figure 12).

Sourdoughs B2 and B3 are rather dominated by the genus *Kazachstania bulderi* (B2) and *Kazachstania servazzii* (B3), species present in homemade sourdoughs. Regardless of the number of refreshes, one dominant yeast species, ≥ 50% of the isolates, seems to be present in all yeasts.

For lactic acid bacteria, the predominance of the genus *Lactobacillus*, observed in Figure 12, is confirmed by identification at the species level by cultural methods (Table IX and Figure 14). This lactic flora appears to be quite diverse for B1 and B2 (large number of species > to 6 and no species > to 50% of the isolates), while B3 and B4, with more refreshments, show a dominant flora at more than 70% (L. *sakei for* B3 and L. *plantarum for* B4). On the other hand, the lactic flora of B5 sourdough (Lab) remains diversified despite the high number of refreshes (13).

A link between the presence of *Lactobacillus sakei* and the practice of chilling between cooler (Sourdough B3) and lower temperatures (Sourdough B1) could be made (Table IX). Sourdoughs B1 and B5 have, in cumulative dominance, an obligatory heterofermentative flora, whereas they are the most different from the 5 batches of Sourdoughs in number of refreshments.

Table IX: Yeasts and lactic acid bacteria of all the experimental yeasts and homemade yeasts of the various experimental bakers (B1, B2, B3, B4 and Lab Levain) identified by cultural method

(Sourdough Exp.: Experimental sourdoughs elaborated in the bakery with different flours)
(Homemade sourdough: Sourdough used by the baker for his own production)
(Abbreviations / Yeasts: *K.: Kazachstania, Wi. : Wickerhamomyces, S. : Saccharomyces, T. Torulaspora*)
(Abbreviations / Lactic Bacteria: *L. Lactobacillus, Ln. : Leuconostoc, P. : Pediococcus, W. : Weissella*)
(Bold or not, underlined or not: fermentative type of lactic acid bacteria)

Bakers (N Refreshments)	Sourdoughs	Yeast species	Lactic acid bacteria species
			Mandatory heterofermentative species Optional heterofermentative species <u>Homofermentative species</u>
B1 (N Raf. 5)	Sourdough Exp.	*S. cerevisiae* (64%), *K. servazzii* (19%), *Wi. anomalus* (17%)	*L. curvatus* (35%), **Ln. mesenteroides** (19%), L. sakei (19%), L. **heilongjiangensis** (15%), **Ln. citreum** (6%), <u>P. pentosaceus </u>(6%)
	Homemade Sourdough	*S. cerevisiae* (100%)	**L. sanfranciscensis** (80%), L. sakei (20%)
B2 (N Raf. 7)	Sourdough Exp.	*K. bulderi* (50%), *Wi. anomalus* (50%)	L. plantarum (44%), L. paralimentarius (28%), L. koreensis (15%), <u>P. parvulus </u>4%, L. **brevis** 3%, L. **heilongjiangensis** (2%), L. hammesii (2%), **Ln. mesenteroides** (2%)
	Homemade Sourdough	*K. bulderi* (100%)	L. plantarum (83%), **L. sanfranciscensis** (17%)
B3 (N Raf. 9)	Sourdough Exp.	*K. servazzii* (65%), *S. cerevisiae* (35%)	L. sakei (76%), L. **heilongjiangensis** (16%), **Ln. citreum** (7%), **Weissella sp.** (1%)
	Homemade Sourdough	*S. cerevisiae* (56%), *K. servazzii* (44%)	L. sakei (50%), **L. heilongjiangensis** (40%), **Weissella sp.** (10%)
B4 (N Raf. 10)	Sourdough Exp.	*S. cerevisiae* (100%)	L. plantarum (87%), L. **brevis** (11%), L. paralimentarius (2%)
	Homemade Sourdough	*S. cerevisiae* (100%)	L. paralimentarius (67%), L. plantarum (33%)
Sourdough B5 Labo (N Raf. 13)	Levain Labo	*S. cerevisiae* (80%), *Wi. anomalus* (10%), *K. servazzii* (6%), *T. delbrueckii* (4%)	**Ln. mesenteroides** (26%), **L. sanfranciscensis** (24%), **L. spicheri** (20%), <u>P. pentosaceus </u>(14%), **Ln. citreum** (13%), **L. rossiae** (3%)

31

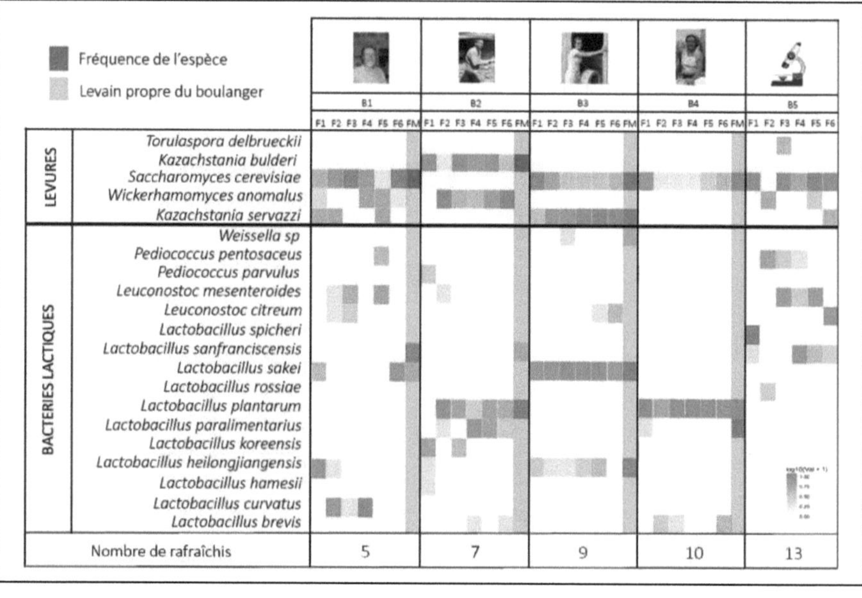

Figure 14: Dominant yeast and bacterial species of all the experimental sourdoughs produced on the different flours (F1 to F6) and the homemade sourdoughs of the different experimental bakers (B1, B2, B3, B4 and B5-Levain Lab). Identification by cultivation method.

No effect of flours on the microflora of sourdough prepared by bakers is observed (Figure 14). Under laboratory conditions (B5), it appears that the microbial flora remain more variable between flours, even though the number of refreshes was high. A low number of refreshes on the one hand (B1) and "aseptic" conditions on the other (B5) could lead to a more marked expression of microbial diversity in the sourdough. These two observations confirm the impact of the environment and baking practices on the microbial flora.

4.2.3. Acidity and pH of experimental starter cultures

Measurements of pH and total titratable acidity of yeast are shown in Figures 15 a) and b).

> **Acidity and pH**
>
> Total titratable acidity (TTA) is the neutralization of the total acidity of a quantity of product (10g of sourdough or bread, in solution in water) by a volume of soda (N/10 in the case of sourdough and sourdough bread). The measurement of the titratable acidity will therefore reflect the actual acid concentration of the product, regardless of the characteristics of the acids and the medium.
>
> The Hydrogen Potential (pH) is an index associating the concentration and the chemical activity of the hydrogen ion H+ in a solution or a hydrated medium, released, for example in the paste, by organic acids. If the chemical activity is strong, this corresponds to a significant dissociation of the H+ ions and the pH value on a scale of 1 to 14, will be between 1 and 7, the value 7 being the neutral point between the acidic and basic character (pH above 7). This is a simple measurement, carried out with a pH meter equipped with an electrochemical electrode.

Figures 15 a and 15 b: Variation in pH (Fig. 13a) and titratable acidity (Fig. 13 b) of experimental starter cultures

(TTA: Total Titratable Acidity: mL NaOH N/10 per 10 g of sourdough)

(B1, B2, B3, B4: Bakers' yeast; F1, F2, F3, F4, F5, F6: Bakers' yeast prepared with the 6 flours; LM: Bakers' home-made yeast). The average values of the 6 yeasts for each baker as well as the values of the home-made sourdough are shown in darker colors following those of the yeasts on the 6 flours.

The range of variation in pH of the yeast is relatively small from 3.86 to 4.15 (Figure 15 a). The pH is related to the baker's practices. In particular, the pH values of the different sourdoughs seem to homogenize with the number of refreshments. The differential (ΔpH) between pH max and pH min, observed for a baker between his different sourdoughs, decreases with the number N of refreshes (B1: N Raf. =5, ΔpH = 0.23; B2 : N Raf. =7, ΔpH =0.14; B3: N Raf. =9, ΔpH = 0.09; B4: N Raf. =10, ΔpH = 0.06).

There is no flour effect on the pH value of the yeast. (Modern mixes average pH=3.95±0.08; Old mixes average pH=4± 0.03). Differences in protein content

(Table VII), and thus potentially in buffering capacity and nutritional resources, therefore do not seem to have any effect on this activity marker.

The total titratable acidity (TTA) of the yeast (Figure 15 b) is between 9 and 13.4 mL NaOH 0.1N / 10 g. This range of variation is related to the high values for baker 1, probably related to the low number of refreshments (5) that would lead to an accumulation of acidity. On the other hand, the lab sourdoughs (data not shown here), with a large number of refreshments (13), have a high average ATT of 13.5 mL NaOH 0.1N/10 g. There is no apparent link between the microbial species, especially lactic acid species, of the different starter cultures (Table IX) and these acidity values, except that the least acidic starter cultures (B2, B3, B4) have a predominantly heterofermentative facultative lactic flora. Note that the B1 yeasts, despite higher acidity, have a relatively less acidic pH; this observation may be related to a higher acetic acid (weak acid) content for these yeasts, as observed on the breads (Figure 24).

There is little or no significant effect of the flour variety. The ATT is slightly lower on average for Modern Blends (10.14±0.84) compared to Old Blends (10.94±0.56).

5. Influence of diversity Wheat/flour/leavening in the baking process

5.1 Breadmaking test protocol

On receipt of the S3 sourdoughs from the bakers, all the sourdoughs are conducted in the same way. The chef sourdoughs are prepared with a first refresh made at 85% hydration (flour base) and with 4 hours of fermentation at 27°C, followed by storage at 4°C for all the sourdoughs. Bread-making trials were then carried out by flour variety over 3 days (D1: F1 and F2, D2: F3 and F4, D3: F5 and F6).

The day before the day of bread-making, each master sourdough is taken again to prepare the all-point sourdough, with 62% hydration, 4 hours fermentation at 27°C and storage at 4°C. The all-point sourdough is re-fermented 2 hours before the start of baking. The flour used is the same as that used for the preparation of the various sourdoughs.

5.1.1. Manufacturing diagram

The protocol was chosen (Table X) based on the experience of the WheatCountry contract, acquired in the trials of sourdough-based breadmaking with ancient and modern population varieties (Triptolème, 2017-3).

Table X: Reference protocol for baking trials

Average temperature of the ingredients	All-point yeast	6 °C
	Flour	20 °C
	Fournil	20 °C
	Water	19 °C
	Base temperature	75 °C
Kneading on spiral kneader	Milling: speed 100 rpm	4min
	Hydration	Consistency function
	Kneading: speed 200 rpm	2 to 4 min in Smoothing function
Tank pointing	Chamber at 27 °C	120 min flap at 60 min
Division	Manual / dough pieces	10 min Mass 500 g
Boulage	Manual	5 min
Relaxation		20 min
Shaping	Manual / Length	15 min / 30 cm
Primer	Chamber at 27 °C	120 min
Cooking	Hearth furnace at 245 °C	30 min

5.1.2. Basic manufacturing formula

The revenues implemented are detailed in Table XI. The elaboration of a baking dough is done by hydration.

Hydration

In baking, as in pastry making, variation in the order of incorporation of ingredients or a group of ingredients (solid and liquid) in a mix is both a technological and practical reality.

The elaboration of a bakery dough is done by hydration of the flour, that is to say :

1. by seeking a constant consistency, a reference for the baker, during the kneading operation, when the quality of the flour is not known :

- By adding water to the flour to make a dough of a given consistency (method for breadmaking tests), the amount of water associated with the solid matter (flour hydration) is deducted. It is expressed in mass or volume of liquid/solid (L/kg) or in proportion of liquid to flour (% water/flour);

- by mixing flour with water to make a dough of a given consistency ("flour for dilution", a term not often used). It is expressed as mass or volume of solid/liquid (kg/L) or as the proportion of flour to liquid (% flour/water).

2. at constant hydration (constant proportion of water to flour), when the quality characteristics of the flour, related to water absorption, are under control (protein content, damaged starches, and non-cellulosic fibres and water content and rheological measurements for consistency determination). There are, however, risks of consistency variation that will have to be corrected.

In the different approaches, the aim is to approach a constant dough consistency, which remains very much linked to the baker's observation; this may vary according to the baker and the process. It can be weakly or strongly hydrated (low or high hydration) if the consistency is strong or weak (hard, firm, soft...).

The determination of bread yield and the search for constant quality leads bakers, especially industrial bakers, to monitor the water content and its evolution during the bread-making process. This is expressed either as the proportion of water in relation to the "as is" material (%/MTQ) or in relation to the dry material (%/MS).

		% sur farine mise en œuvre	masses (g)	% sur pétrissée	% sur farine totale
Formule du levain chef					
farine		100	261	6	9
eau		85	222	5	
Total levain chef		185	483	10	
Formule du levain tout point				0	
farine		100	568	12	20
eau		62	349	8	
Levain chef		85	483	10	
Total levain		247	1400	30	
Formule de la pétrissée ou pétrie				0	
farine		100	2000	43	71
eau		60	1200	26	
eau ajoutée		0	0	0	
sel		2	40	1	
levain tout point		70	1400	30	
Total pétrissée			4640	100	
Formule ingrédients de la pétrissée				0	
total farine		100	2829	61	100
total eau		73,92	1771	38	
sel		1,67	40	1	
Total ingrédients			4640	100	

formule de départ
proportion d'eau par rapport à la farine qui permet d'orienter la consistance du levain
proportion de farine fermentée par rapport à la farine totale qui donne une indication des risques qualitatifs sur le gluten
proportion de farine apte à former du gluten pour assurer la rétention gazeuse
proportion de levain dans la pâte qui donne une indication sur l'activité microbienne de la pâte
Proportion de farine totale utilisée pour une masse de pâte qui permet d'envisager le besoin journalier, hebdomadaire et annuel
Quantité de pâte à pétrir en fonction du nombre et du poids des pains fabriques

Sourdough accounts for 70% of the flour or 30% of the total dough, close to the average baker's practice.

Sourdough production was initially carried out by bakers (Table VIII) with a hydration (water content) of between 70 and 100% of the flour, with an average of 85% being used in the production of sourdough (Table XI).

5.1.3. Rating grid

The observations made during the breadmaking trial are based on a sensory methodology whose terms and the way of characterizing and noting have been defined in the Triptolème documents (feuille d'expérimentation boulange, 2014 and glossary expérimentation boulange, 2014) and are also based on the bases of the breadmaking trial in the AFNOR methodology (NF VO3-716).

Caractéristiques de la pâte	insuffisance				excès		
Interprétations, observations et notes	1	4	7	10	7	4	1
Rapidité lissage			X				
Collant de la pâte					X		
Consistance				X			
Extensibilité		X					
Elasticité	X						
Relachement					X		
PETRISSAGE							
Pousse en cuve		X					
Détente relachement						X	
POINTAGE							
Allongement							X
Déchirement						X	
Elasticité	X						
Collant de la pâte						X	
FACONNAGE							
Activité fermentaire							
Déchirement					X		
APPRET							
Collant de la pâte				X			
Tenue		X					
MISE AU FOUR							

Caractéristiques du pain	insuffisance				excès		
Interprétations, observations et notes	1	4	7	10	7	4	1
Section		X					
Couleur					X		
Epaisseur				X			
Croustillant		X					
Coup : Développement	X	X					
de : Régularité				X			
lame : Déchirement					X		
ASPECT DU PAIN							
Couleur				X			
Texture : Souplesse			X				
Elasticité				X			
Collant				X			
Alvéolage : Régularité				X			
Epaisseur				X	X		
Odeur : Saveur				X			
ASPECT MIE							
Volumes des pains (cm^3)							

Figure 16: AFNOR-type scoring grid used for the scoring of breadmaking trials (example of the scoring of leaven B1 and flour F3)

5.2. Behaviour of dough during baking

5.2.1. Observations on the conduct of baking

Given the lack of preliminary trials, the kneading was conducted with a correction of hydration according to the consistency at the milling stage and the appearance of smoothing during kneading. The kneading time was higher for the F4 and F6 flours to obtain the beginning of dough smoothing. The dough temperatures obtained between 24 °C and 25.5 °C are on average higher for the Modern Mixes (approx. 1 °C); this can be explained by 3 factors, a higher average kneading time, more elastic dough and higher vat temperatures because the Modern Mixes were systematically baked after the Ancient Mixes series. This difference is nevertheless within the range mentioned in the AFNOR bread-making test which is 25 °C ± 1 °C. In spite of different kneading times and the variation in dough temperatures, no significant differences in consistency at the end of the kneading process can be observed.

5.2.2. Physical characteristics of the doughs being baked

During the bread-making tests, several observations are made, in particular: hydration, elasticity, strength; they are good indicators of the visco-elastic properties of the doughs and their evolution during the bread-making process.

Table XII: Hydration of dough during kneading (values in % relative to flour)

Flours (MM: Modern Mix, MA Old Mix in gray)	Bakers				Medium / Flour
	B1	B2	B3	B4	± Deviation Type
F 1	58	59	60	60	59,3 ±0,8
F 2	60	64	64	63	62,8 ±1,6
F 3	58	59	61	59	59,3 ±1,1
F 4	62	64	66	64	64,0 ±1,4
F 5	58	60	63	60	60,3 ±1,8
F 6	65	68	70	67	67,5 ±1,8
Average / Baker	60,2	62,3	64,0	62,2	MA: 59.6 ±0.5
± Deviation Type	±2,6	±3,3	±3,3	±2,8	MM: 64.8 ±2.0

The hydration (Table XII), obtained by manual observation of the consistency at the beginning of kneading, related to the viscous properties, is lower with the Ancient Mixes. They are mainly explained by lower levels of damaged starches than in Modern Mixes (Table VI), despite higher protein and gluten levels. Knowledge of the fiber content of flours would make it possible to refine this analysis.

From the rheological point of view, the results (Figure 17) on relative hydration differences obtained with Mixolab provide a prediction ($R2 = 0.70$) of hydration in baking. This is even if, due to the principle of the method, the Mixolab data integrate both viscous and elastic properties during kneading and not, as in baking practices, the evaluation during kneading stoppage.

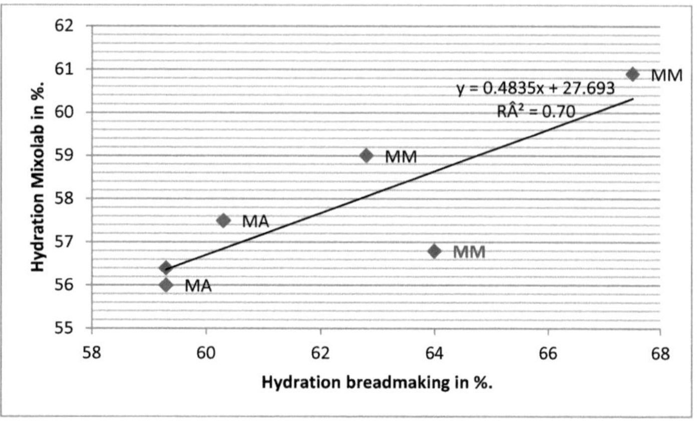

Figure 17: Relationship between the hydration level to obtain a constant Mixolab resistance torque and hydration by observation in baking (MA Ancient Mix; MM Modern Mix)

In Figure 17, the MM point colored red, corresponding to flour F4, has a peculiarity; it is located in the modern block by its damaged starch content (Table VI) and close to the Ancient Mixes by the resistant couple C2, lower, in the Mixolab (Table VII). This lower resistance may explain, for a constant C1 couple, a reduced hydration for F4, compared to the other Modern Mixtures F2 and F6. It may explain, in part, its shifted positioning with respect to the regression line.

The evolution of elastic characteristics during baking is an important criterion for predicting the development capacities in fermentation and baking; it is related both to the intrinsic characteristics of the flour and to the baker's handling of the baking process.

By manually observing the elasticity of the dough at the end of the kneading process, the elasticity values of the Ancient Mixes are, on average, lower than those of the Modern Mixes (Figure 18). These results are related to the state of the gluten assessed by physico-chemical methods (Table VII). The elasticity values, also for F4 flour, reflect the results of the Gluten Index, which are higher for the Modern Mixes.

Doughs prepared with B4 sourdoughs, which are considered more elastic at the end of kneading (Figure 18), are stronger than other doughs at pointing, their elongation is more difficult to shape, and the tearing and holding of the dough at the primer (Figure 19) is greater than other bakers' sourdoughs.

Certain hypotheses can be put forward to explain the higher level of elasticity and PTO for B4 yeasts:

- a state of the gluten less degraded at the origin in the sourdough chef due to the greater number of refreshes ;

- less effect of acidity on the state of cohesion of the gluten; acidity is lower for B4 yeast compared to B1, B2, B3 yeast from bakers (Figures 15 b);

- an oxidizing potential of the leaven which would be higher due to the higher number of refreshes.

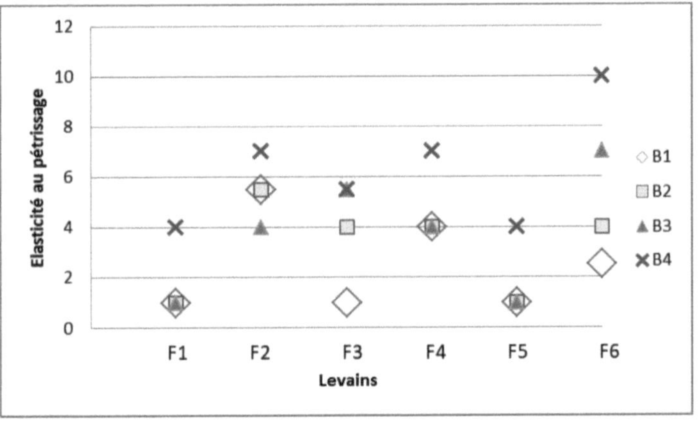

Figure 18: Influence of sourdough and flour on the elasticity of dough during kneading.
(Ancient Blends F1, F3, F5; Modern Blends F2, F4, F6)

The oven resistance does not systematically reflect the elasticity of the dough, which is also associated with the retention and gaseous production properties of the dough and its degree of oxidation, linked to the power take-off. Nevertheless, if we compare bakers B4 and B1, there is a correspondence between the decrease in elasticity and dough behaviour (Figures 18 and 19), and a very clear difference in the impact of the effect of the yeast of these two bakers on these two dough descriptors. For bakers B2 and B3, it is difficult to show a strong link between dough elasticity and dough strength and to show a specific impact of the yeast of these two bakers on these characteristics.

Figure 18 shows that the persistence of the elastic characteristics identified in the flours is more marked with Modern Mixes than with Ancient Mixes, which may mean that the two types of wheat are not affected differently by the yeast.

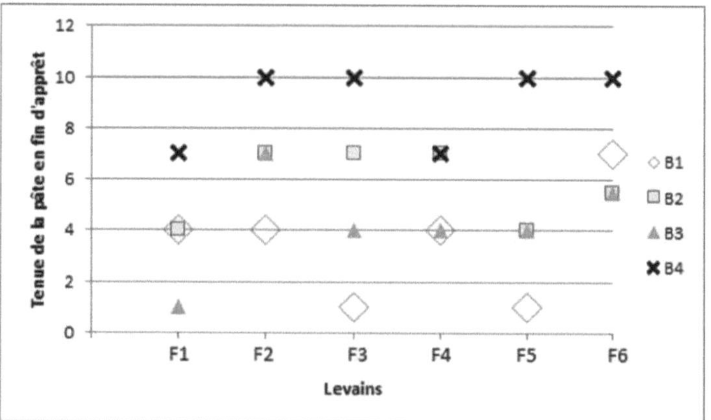

Figure 19: Influence of sourdough and flours on the behaviour of pasta at the end of the preparation process

5.2.3. Methodological approach to characterize proteins during bread-making process

The objective of this approach, which was not directly integrated into the Bakery project, was the search for indicators that would make it possible to highlight the evolution of proteins during bread-making and/or the impact of sourdough on the state of the proteins. The traditional measurements used in the cereal sector do not make it possible to qualify the state of proteins during bread-making. A first methodological approach, based on a partial sampling of doughs from different sourdoughs and flours, was carried out by measuring the protein fractions in HPLC-type chromatography. Their quantification at different stages of production (Figure 20) revealed some trends, namely: higher F1 and F2 fractions of high molecular weight (glutenins) at the end of preparation and scoring compared to the end of kneading for the lower molecular weight fractions higher on F3 and F4 (gliadins) and F5 (small proteins). We can hypothesize re-polymerization effects of the proteins during baking compared to kneading that can be explained by the oxidation of the proteins. In the trials carried out, different developments also appear depending on the sourdoughs studied by some bakers and the type of wheat. The phenomenon of hydrolysis, the consequences of which would result in an increase in the smaller protein fractions between the end of kneading and the end of preparation, has not been identified. In the future, for this type of study on the effect of sourdough in baking, the monitoring of these indicators seems interesting and relevant to better understand their effects, including the oxidation-reduction reactions and the links with the "dough power take-off" observed by bakers.

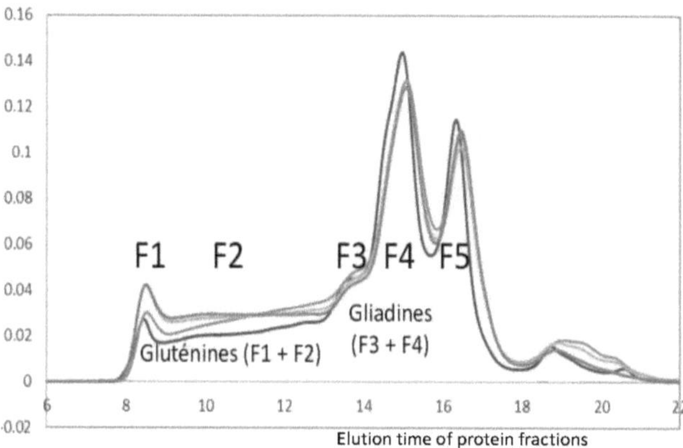

Figure 20: Protein profile at different stages of baking of a dough made of F4 flour and leaven by baker B4

Mix Flour + Sourdough all point :
Dough End of kneading :
Paste at the end of pointing :
Paste at the end of primer :

5.2.4. Monitoring the growth of the dough during the baking process

The **growth is the** result of several parameters that reflect the level of development of the dough during and at the end of finishing. It is correlated with the production and gas retention and the elasticity/extensibility balance of the dough. In bakery, it can be attributed additional qualifiers linked to development characteristics, either to the speed of development (fast or slow growth) or to the shape of the dough during development (round or flat growth). The growth also varies according to the atmospheric pressure.

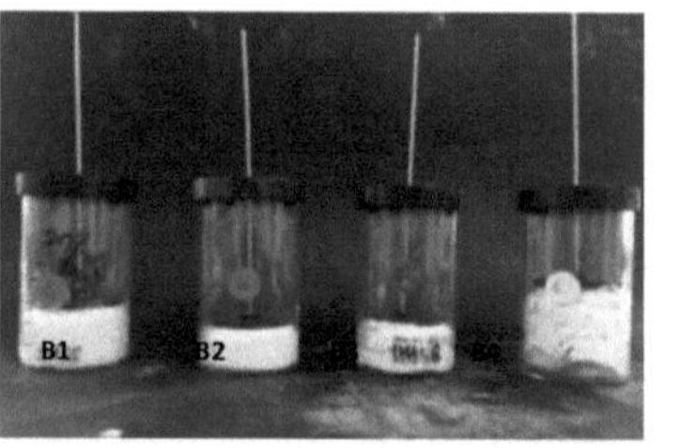

Figure 21: Visualization of the maximum development of the baked dough with flour F6 on the basis of the 4 yeasts (from left to right from B1 to B4) in the growth meter.

A sourdough effect is observed (Figure 22a); the sourdoughs from bakers B1 and B4 show higher growth than B2 and B3. The heights of development obtained with B4 and B1 sourdoughs compared to B2 and B3 seem to benefit from a higher gas production, which can be related to the density of *Saccharomyces cerevisiae* type yeasts (Table IX).

Note that pasta prepared with B4 yeast has a shoot showing good gas retention. The higher microbial activity of these yeasts (Figure 13) does not therefore lead to greater hydrolysis actions.

As regards flours (Figure 22b), the standard deviations observed do not allow the conclusion to be drawn that there is a significant difference in pasta growth as a function of the Ancient and Modern mixes, although a higher trend is observed for the Modern than for the Ancient mixes up to T90 min. The characteristics of proteins and gluten, their structuring in baking and the fermentation activities of the Old and Modern Mixes do not seem to have an impact on the growth.

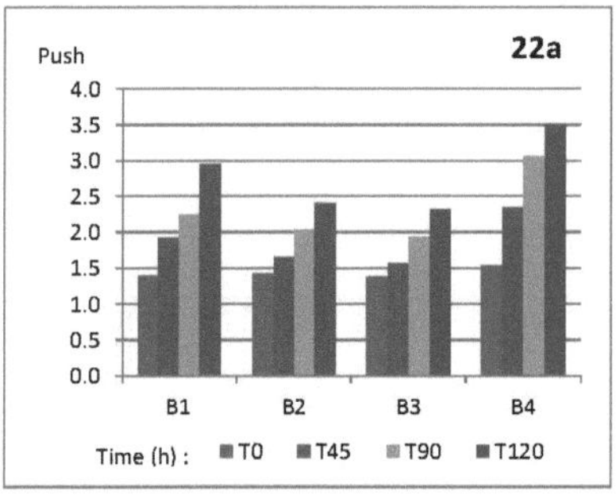

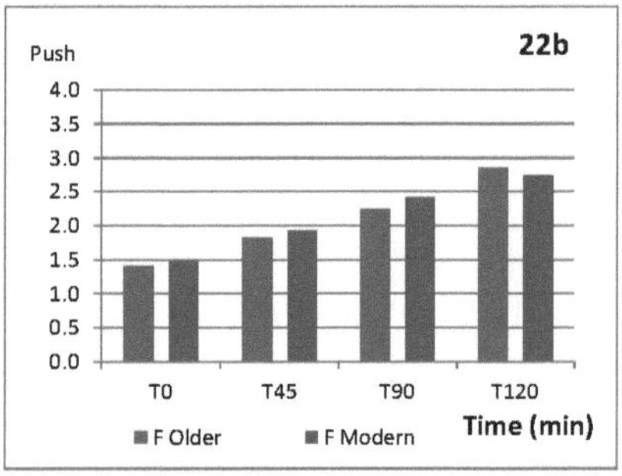

Baker	T0	T45	T90	T120
B1	1,4±0,1	1,9±0,2	2,3±0,3	3,0±0,2
B2	1,4±0,1	1,7±0,1	2,0±0,2	2,4±0,5
B3	1,4±0,1	1,6±0,2	1,9±0,1	2,3±0,5
B4	1,5±0,1	2,3±0,2	3,1±0,8	3,5±0,6

Flours	T0	T45	T90	T120
Average MA	1,4±0,04	1,8±0,01	2,2±0,17	2,9±0,21
Average MM	1,5±0,05	1,9±0,05	2,4±0,32	2,7±0,26

Figures 22 a and 22 b and associated tables: Growth measurements during bread-making. Mean Sourdough values for the various bakers B1, B2, B3, B4 (22 a) and for the old MA and modern MM flours (22 b).

5.2.5. Follow-up of the gaseous production during the bread-making process

The **gas production** can be measured by the pressure variation in a closed system, here Ankom device (Figure 23), in order to take into account both the gas retained in the dough and the gas released in the headspace. It corresponds theoretically to the measurement of all the carbon dioxide produced in the paste, provided that the difference in permeability and resistance of the pastes to internal pressure is neglected. This measurement therefore makes it possible to evaluate, in part, the activity potential of a ferment, in this case yeast, on a given flour, independently of the gas retention specific to this flour. It also makes it possible to evaluate the potential of a given flour to provide substrates or favourable conditions for the yeast.

Figure 23: Measurement of the apparentgaseous production of pasta (Ankom System)

The apparent gas production was monitored as soon as the kneading was finished and the results obtained for the 24 doughs prepared on the different flours with the 6 head sourdoughs from each of the 4 bakers were processed by means of the 6 kinetics for the 4 bakers on the one hand (Figure 24) and by category of flour, modern and old, on the other hand (Figure 25).

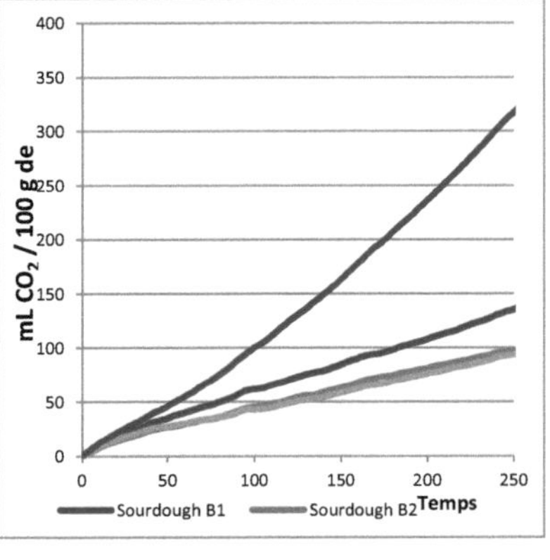

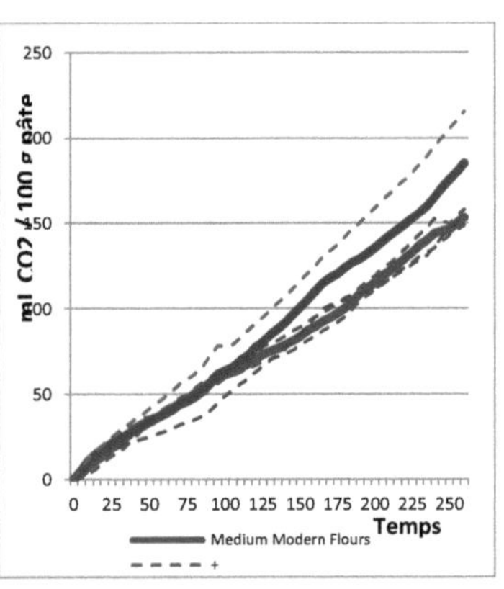

Figure 24: Apparent gas production kinetics during baking by baker (Ankom meter; average of the 6 sourdoughs elaborated on the 6 flours for each baker)

Figure 25: Apparent gaseous production kinetics during baking by flour category (Ankom meter; modern flours F2, F4, F6 and old flours F1, F3, F5; average between the 4 bakers, i.e. 12 doughs per flour category)

With regard to the sourdough effect (Figure 24), the sourdoughs from bakery B4 have a significantly higher activity than the other sourdoughs. Sourdough B1 also stands out from the other two groups of sourdoughs from bakeries B2 and B3. The flours of these two groups of sourdoughs, B4 and B1, are microbiologically characterized by the presence of strains of *S. cerevisiae* (Table IX). In addition, these two groups have significantly higher microbial densities of yeast on average (B4: $1.^{108}$ and B1: $6.^{107}$ CFU/g) than the other yeasts (B2: $3.^{107}$ and B3 $1.4.^{106}$ CFU/g). It therefore seems that, under the experimental conditions, the density of the yeasts, the presence of this yeast species and the associated lactic flora favours gas production.

When all flours are considered (data not shown), no significant effect of flour on gas production was found. Comparing the results obtained for the Modern Mixes flours and the Old Mixes flours (Figure 25), a slightly higher gas production appears with the latter; however, in view of the variations observed, this difference is not statistically significant.

The absence of a flour effect on gas production despite the difference in the level of damaged starches (Table VI: 21.1 UCD on average for MA versus 24.2 for MM) between Modern and Ancient Blends would require a more refined analysis.

Since the amylase activities were not significantly different among the flours in the experiment (Table VII), it is possible to hypothesize that the quantity of carbohydrate substrates (sugars) available to the yeasts is not limiting and therefore does not show any flour impact on this point.

6. Analysis of breads obtained from experimental leavens

6.1. Alveolar structure of breads

The honeycombing of the crumb and the crust of the bread is defined by :
- the number of cells per unit area or volume
- the average size of the alveoli
- the dimensional regularity of the cells
These characteristics can be associated with the alveolar density. A high alveolar density is characterized by a high number of alveoli which leads, for a surface or a volume of bread, to smaller and more regular alveoli. The thickness of the alveolar walls decreases accordingly. The alveolar structure is created during kneading but evolves during bread-making, the alveoli deform if there is enough gas produced, they increase in number during the dough flaps and shaping, they decrease in number, by the phenomenon of coalescence, if the stability of the dough is not sufficient (very hydrated dough that lacks consistency, too strong enzymatic hydrolysis activities).

B1 to B4 are shown in Table XIII.

Table XIII: Slices of bread from the different trials scanned
(Sourdoughs B1, B2, B3, B4 baked with Flours F1 to F6; see Table III and IV)

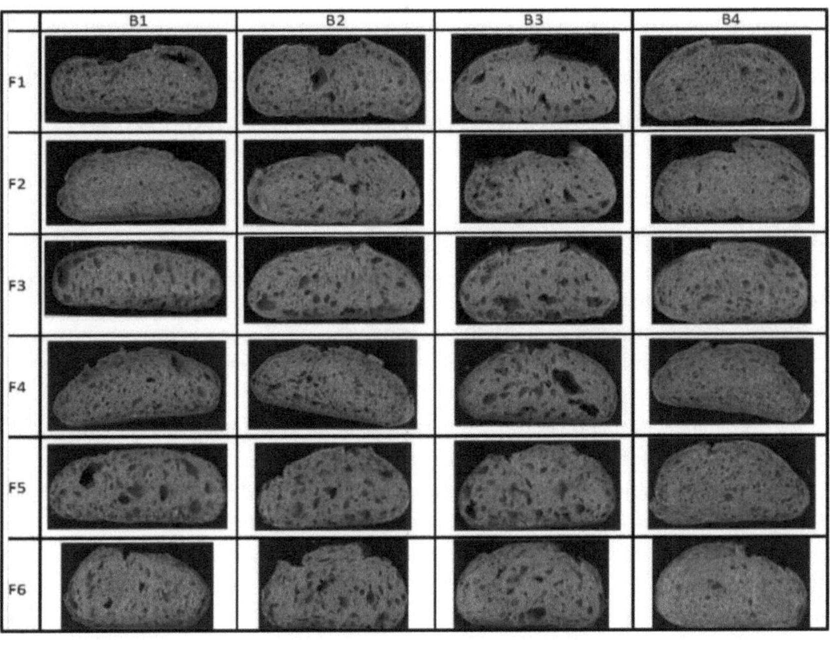

Since imaging treatment of the alveolar structure could not be carried out, it is difficult to quantify the number, regularity and size of the alveoli and the thickness of the alveolar walls. The observation of the images (Table XIII) nevertheless makes it possible to affirm that the bread crumbs of the baker B4 are more "foamy", i.e. the number of cells per unit area and the regularity of the cells are higher. The size of the alveoli is also lower in B1, B2 bakers and this is even more marked compared to B3 baker. A link can be made with the number and activity of the yeasts (for example: N yeasts for B3 the weakest), but this observation is not sufficient in itself. Indeed, the characteristics of alveolar regularity are associated in bakery technology with gas production, gas retention and the stability of the gluten network. When these factors are more marked, the stability of the alveoli is better, the coalescence phenomenon is less and consequently the integrity of the alveoli is better preserved.

The flours of the Old Mixes F1, F3, F5 seem to have a more irregular honeycomb pattern compared to the flours of the Modern Mixes F2, F4, F6, which remain "tighter". This effect is particularly visible in B1 and B4.

6.2. Bread volume

Factors in the volume of a loaf of bread

To ensure aeration and development, the dough must :

- contain gas (role of aeration during kneading and gas production during fermentation);

- retaining gas: role of gluten, from its structuring to kneading and its stabilization by the oxidation-reduction actions during the bread-making process;

- Expansion (role of fermentation activity, of the good elasticity/extensibility balance of the gluteal network and of gas expansion at the beginning of cooking);

- stabilize this expansion during cooking by the gelatinization of starch and the coagulation of gluten.

The main factors for obtaining these results are the wheat proteins, the quality of the gluten formed, the fermentation activity, the stability of the dough, which depends mainly on the baker's working methods, and the intensity of the enzymatic actions of the flour and the microorganisms (hydrolysis and oxidation-reduction) during the baking process.

The volumes of the loaves were measured (Table XIV) by displacement of round seeds, of low density not too high (rapeseed, millet, mustard...) in a cylindrical container specific for this measurement (bread volumeter). The same measurement is made with a control wedge, of length and volume close to the bread measurements, and whose volume is precisely known. By difference, between the volumes of displaced seeds, we can deduce the volume of the bread (AFNOR V03-716 standard). Moreover, breads are the subject of a section measurement, by the ratio, measured in centimeters, between the height and the base (Table XV).

Table XIV: Bread volume in cm3 according to flours and bakers
(MM: Modern Mix, MA Ancient Mix) (Old flours in gray)

Flour	Baker				Average
	B1	B2	B3	B4	
F1	1016	900	774	1106	949
F2	1005	847	818	1145	954
F3	1023	1010	953	1479	1116
F4	955	1026	868	1186	1009
F5	920	888	802	1116	932
F6	1102	914	957	1368	1085
Average	1004 ±57,1	931 ±65,1	862 ±71,4	1233 ±140,5	MA: 999 ±83.0 MM: 1016 ±53.7

The average volume of the Modern Mixtures is very much the same as that of the Ancient Mixtures, 17 cm3 more (Table XIV); this confirms the observations of growth measurements (Figure 20b). However, for the same volume, the bread section is different and on average the Ancient Mixes have flatter bread sections (Table XV), which is consistent with the dough's behaviour before baking (Figure 18). In the Modern Mixes, only F4 is similar to the Ancient Mixes, and the same was observed with the Mixolab (§ 5.2.2).

The influence of the origin of the yeast is very marked, with a maximum difference of about 370 cm3 depending on the yeast. Bread volumes are higher with B4 and B1 sourdoughs, which is consistent with the observations of higher shoot height (Figure 20a) and higher gas production (Figure 22) for these sourdoughs.

If we consider the section of breads (Table XV), it is different according to the yeasts; the breads resulting from the yeasts of baker B1 are flatter than those of baker B4 in particular. Here again, there is consistency with the observations, during kneading, on the lower dough holding for B1 (Figure 19).

Table XV: Section of breads according to sourdough and flours
(Insufficient values according to the bread-making monitoring grid in figure 16).

Flours	Baker				Medium / Flour
	B1	B2	B3	B4	
F1	4	4	4	7	4,8 ±1,5
F2	4	7	7	7	6,3 ±1,5
F3	1	5,5	4	10	5,1 ±3,8
F4	1	7	4	7	4,8 ±2,9
F5	1	4	4	7	4,0 ±2,4
F6	7	7	7	10	7,8 ±1,5
Average / Sourdough	3,0 ±2,4	5,8 ±1,5	5,0 ±1,5	8,0 ±1,5	MA: 4.6 ±0.46 MM: 6.3 ±1.22

6.3. Acidity and pH of the breads

Tables XVI a) and b) show the average pH and total titratable acidity (TTA) values of breads obtained with the different types of sourdough depending on the flours and baker's yeast.

Measurements of pH and total titratable acidity of yeast are shown in Figures 15 a) and b).

The pH results of the breads (Table XVI a) on the 6 flours confirm that all the breads cannot be classified as "French traditional sourdough bread" with a maximum pH base of 4.3 (French Decree on Bread 93...). Only bakers B2 and B3 could claim this appellation provided they meet the criterion of 900 ppm minimum acetic acid.

Table XVI a) and b): pH and ATT of breads according to flours and bakers (MM: Modern Mix, MA Ancient Mix) (Old flours in gray)

a))

Flour	pH					ATT				
	B1	B2	B3	B4	Average	B1	B2	B3	B4	Average
F1	4,61	4,24	4,22	4,39	4,36±0,16	7,39	9,17	8,99	8,48	8,51±0,7
F2	5,18	4,32	4,23	4,57	4,57±0,37	5,59	7,55	8,77	7,37	7,32±1,1
F3	4,92	4,28	4,24	4,40	4,46±0,27	6,97	7,93	7,89	6,99	7,45±0,5
F4	4,29	4,27	4,14	4,44	4,29±0,11	8,21	7,72	8,83	7,03	7,95±0,6
F5	5,05	4,31	4,18	4,51	4,51±0,33	6,61	8,79	9,16	8,37	8,23±1,0
F6	4,31	4,20	4,15	4,44	4,28±0,11	8,89	9,24	9,91	7,4	8,86±0,9
Avg.	4,73 ±0,35	4,27 ±0,04	4,19 ±0,04	4,46 ±0,06	MA: 4.44 ±0.06 MM: 4.38 ±0.13	7,28 ±1,1	8,40 ±0,7	8,93 ±0,6	7,61 ±0,6	MM: 8.06 ±0.45 MM: 8.04 ±0.6

Depending on the flours, the average pH ranges from 4.28 for F6 to 4.51 for F5. For bakers, the average pH ranges from 4.19 for B3 to 4.73 for B1. The origin of the sourdough therefore leads to greater variations than the flour factor. Sourdoughs B1 and B4 have the highest pH values.

The values of ATT (Table XVI b) vary globally from 5.6 to 9.9 mL NaOH /g and are in agreement with the pH values. It should be noted that for a given baker, the APTR values of breads are correlated with the APTR of sourdough and that the latter is therefore predictive of the acidity of the breads. B1 breads show the greatest variation in pH and ATT from one flour to another. Breads made from B1 and B4 bakers' yeast are significantly less acidic than breads made from B2 and B3 bakers' yeast. One hypothesis here is probably a competition effect between the yeast microflora and the lactic flora. *Sourdoughs* B1 and B4, with the species *S. cerevisiae in* the majority (Table IX), could see their lactic microflora being less active in terms of acidification.

There are no significant differences between Old and Modern Mixes in either pH or ATT data for individual bakers or for average AM/MM values.

6.4. Maltose content of breads

The content of different carbohydrate compounds (glucose, fructose, maltose, ...) in the breads was analyzed by HPLC chromatography. The results concerning maltose are presented in Table XVII.

Table XVII: Maltose contents (g/kg) measured on breads (MA Ancient Mix, MM Modern Mix) (Ancient flours in gray)

Flour	Baker				Average
	B1	B2	B3	B4	
F1	2,6	3,1	3,4	2,0	2,8 ±0,5
F2	4,2	6,8	5,0	4,0	5,0 ±1,1
F3	3,1	3,9	2,3	1,7	2,8 ±0,8
F4	3,5	6,7	4,5	3,5	4,6 ±1,3
F5	5,5	4,7	3,0	2,9	4,0 ±1,1
F6	3,6	5,5	3,8	2,6	3,9 ±1,0
Average	3,8 ±0,9	5,1 ±1,4	3,7 ±0,9	2,8 ±0,8	MA: 3.2 ±0.6 MM: 4.5 ±0.5

The analysis of the residual maltose of the breads allows us to consider a relation with the release of C02 and yeasts from the yeasts of the leavens. Baker's yeast B4, characterized by high gas production (Figure 23), has the lowest average maltose content. The average obtained with bakers B1, B3 and B4 is lower than the average of the baker's yeast B2. The yeasts from bakers B1 and B4, in which *S. cerevisiae* is dominant (Table IX), have higher yeast activity (Number of yeasts/g, gas production), which may explain the higher consumption of maltose. Baker's yeast B3 also contains the species *S. cerevisiae* (Table IX and Figure 14). The presence of this yeast with a positive maltose phenotype (ability to consume using maltose as a substrate), unlike the genus *Kazachstania sp.* (presence in B2 yeast and also in B3) could therefore be related to the lower residual maltose content.

The results show a lower residual maltose content on average on the Old Blends (F1, F3, F5), than on the Modern Blends (F2, F4, F6): 3.2 ±0.6 g/kg against 4.5 ±0.5 g/kg. This can be explained by the lower content of damaged starches in old flours compared with modern varieties (Table VI), with a slower action of amylases for hydrolysis of the starch chains in the granule. The yeasts consume first the available substrates of the flour, then those resulting from amylase activity, and first glucose and then maltose. Less degradation of the starch will lead to a lower "stock" of fermentable carbohydrates and therefore a lower final residual quantity.

Nevertheless, and in the context of this study, this difference does not translate into an effect of flour variety on sprouting and gas production (Figures 21 and 23), as the content of carbonaceous substrates (glucose, maltose) is probably not limiting for fermentation.

It should be noted, however, that the residual maltose after cooking may be the result of a difference in hydrolysis during cooking, depending on the level of damaged starches and amylase activity.

6.5. Lactic and Acetic Acid Content of Breads and Fermentation Quotient (FQ)

Organic acids: Acids formed by living organisms. Acids are characterized by the presence of at least one COOH function in their chemical formula. The **dissociation,** in a hydrated medium, of the acid group (COOH) into COO- and H+ contributes to increase the concentration of H+ ions in the medium and to decrease the **pH** value (hydrogen potential) of the medium (the higher the concentration of H+ ions, the lower the pH of the medium: inversely proportional relationship). This dissociation capacity is variable according to the acids and reflects the strength of an acid. Organic acids are considered as weak acids, i.e. they dissociate with greater difficulty than a strong acid such as hydrochloric or sulphuric acid. They therefore cause the pH to decrease less. In **yeast, we** find mainly lactic acid and acetic acid produced by yeast microorganisms. Lactic acid dissociates more than acetic acid, so lactic acid is a stronger acid than acetic acid. In addition, each acid has its own sensory characteristics.

The notion of Fermentation Quotient (FQ): Sourdoughs are characterized by the production of lactic and acetic acids, in variable contents and proportions according to the microflora present (Homo / heterofermentative lactic flora, cf. paragraph 1.1), and the conduct of sourdough and bread-making. From the lactic and acetic acid data, it is possible to determine the Fermentation Quotient (FQ: Molar ratio between these two acids). This FQ indicates the more lactic or more acetic tendency of a sourdough bread: a FQ above 5 indicates a fermentation with a lactic tendency, a FQ below 3 shows an acetic tendency. These variations depend both on the nature of the microbial flora of the sourdough, the activity of this flora, the process parameters (time, temperature...).

The organic acid content of the breads (lactic and acetic acids) was analyzed by HPLC (High Pressure Liquid Chromatography). The estimated values in g/kg are shown in Figure 26a and the Fermentation Quotient in Figure 26b.

Lactic acid contents range from 4.3 to 6.9 ± 0.7 mg/kg and are higher for breads made from B2 and B3 yeast, according to the TTA measurements.

The acetic acid contents per baker range on average from 0.7 to 1.5 ± 0.3 mg/kg. These variations, from simple to double, on this acid are stronger than on lactic acid. Breads made with B1 sourdough have the highest estimated average acetic acid values (1.4 ±0.6 mg/kg), while those made with B4 sourdough have the lowest (0.7 ±0.2 mg/kg). This acetic acid content could be penalizing, for B1 yeast, as mentioned above, for the gaseous production by the yeasts.

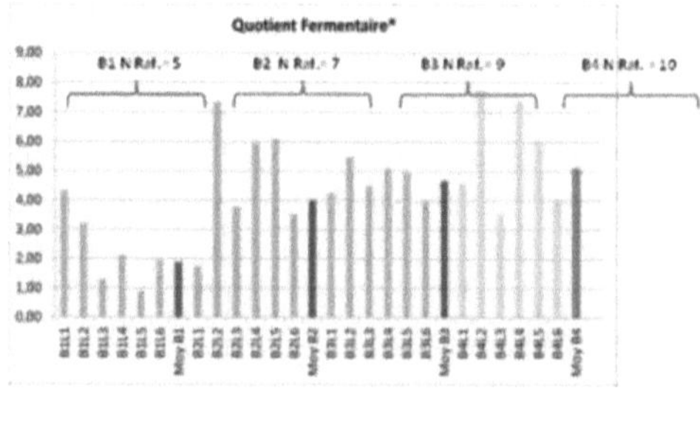
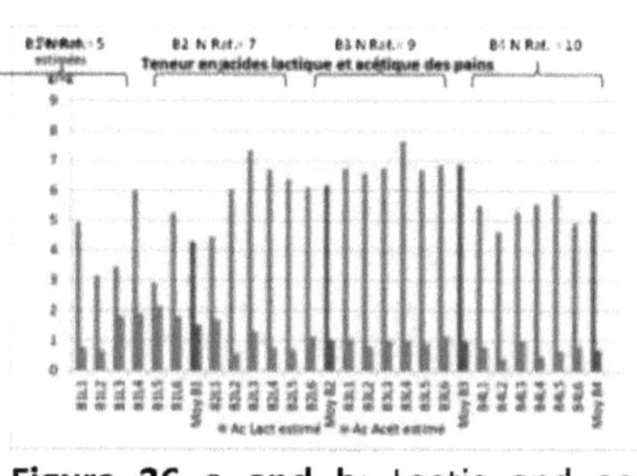

Figure 26 a and b: Lactic and acetic acid content of bread (Figure 25a). Fermentation Quotient (FQ) value of the lactic/acetic acidity of the breads (*FQ: molar ratio of [lactic acid] / [acetic acid]).

The Fermenting Quotas vary from 1 to 7.75 for all breads. B1 yeasts lead to lower FQs, less lactic and more acetic. Sourdough B4 has the highest FQs. It seems that a lower number of refreshes of the yeast over the period considered, leads to a lower FQ (B1: 5 refreshes   B2: 7   B3: 9   fewer heterofermentative acetic acid-producing lactic acid bacteria than the microflora of the yeast of other bakers.

As far as flours are concerned, there are variations within the same bakery, except for B3, which is fairly homogeneous. There is no significant difference in organic acid content between old and modern flours.

Furthermore, the relationship between the acetic acid content of breads and the elasticity scores of doughs during baking shows an inverse relationship between the elasticity of doughs and the acetic acid concentration regardless of the flours (Figure 27).

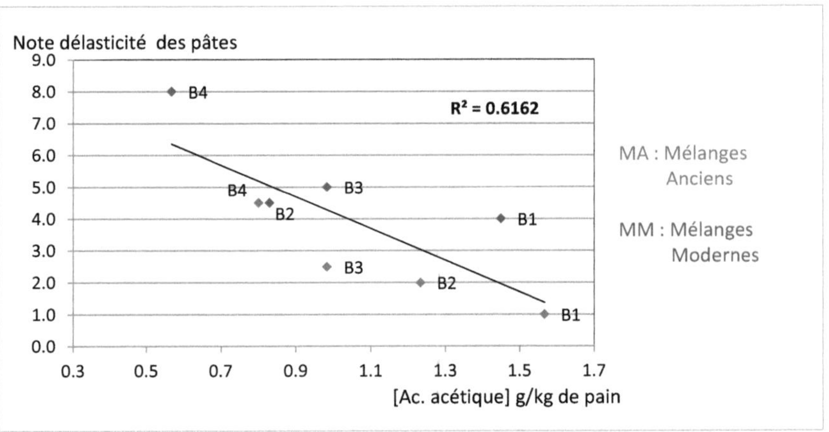

Figure 27: Relationship between acetic acid concentration in breads and dough elasticity

It is as if a lower acetic acid content leads to more elastic pasta. Such a link was not observed either with total titratable acidity or with lactic acid. The acetic acid content of a sourdough is the result of a metabolism that favours this acid over lactic acid, either because of the nature of the lactic bacteria present (obligatory heterofermentative), or according to the parameters of the sourdough (N and frequency of refreshments, time, etc.).

The content of organic acids can have a direct effect on the gluten network by modifying the ionic bonds it generates, thus making the proteins less bound and therefore more solubilizable. Surprisingly, however, this effect seems to be marked by the content of acetic acid, whose concentration and dissociation coefficient (see organic acids insert, § 6.5) are lower compared to lactic acid. In view of these observations, it is possible to make several hypotheses:

- the lower number of refreshes would be an indicator of induced effects, in particular by a modification of the properties of the proteins and their stability, in a more reducing environment (cf. gluten insert § 3.3).

- Stress effects, particularly related to acidity and acetic acid, on the microorganisms could therefore release reducing elements in the dough into the environment.

6.6. Comparison of the biochemical characteristics of breads by baker

More complete biochemical data including metabolites derived from yeast activity (glycerol, ethanol and CO_2 at 4 h) have been taken into account and are presented in Table XVIII.

Table XVIII: Summary of the main physico-chemical characteristics of bread and doughs

(Average of 6 sourdoughs per baker)

Baker	Characteristics of the breads												Pasta
	pH		ATT (mL NaOH /10g)		Lactic Acid (g/Kg)		Acetic Acid (g/Kg)		Fermentation Quotient		Glycerol (g/Kg)		Ethanol (g/Kg)	CO_2 mMol /100g at 4 hours
B1	4,73	5 ±0,3	7,28	±0,07	4,28	7 ±1,1	1,51	±0,58	1,89	±1,17	1,83	±0,27	0,64	4 ±0,5	5,8
B2	4,27	4 ±0,0	8,40	±0,69	6,18	9 ±0,8	1,03	±0,40	4,02	±1,89	0,98	±0,33	0,42	9 ±0,1	4,2
B3	4,19	4 ±0,0	8,93	±0,60	6,88	5 ±0,3	0,98	±0,11	4,67	±0,52	0,18	±0,25	0,23	3 ±0,2	4,0
B4	4,46	6 ±0,0	7,61	±0,60	5,30	2 ±0,4	0,69	±0,20	5,11	±1,62	1,88	±0,24	1,42	3 ±0,5	13,5

It is as if the B1 and B4 yeast favored the yeast activity markers (Glycerol, Ethanol, CO_2 at 4h), thus confirming the observations on the growth and volume of the breads. At the same time, the activity markers of the lactic acid bacteria (lactic acid, TTA and pH), except for acetic acid, show a relatively lower activity of the lactic acid bacteria for these same samples B1 and B4. The two groups of yeast B1 and B4 are characterized by a yeast flora marked by a dominance of *S. cerevisiae of* the positive maltose type. The B2 and B3 yeast groups are characterized by a yeast flora dominated by *Kazachstania sp., a* negative maltose species. It is possible to hypothesize that in the presence of *S. cerevisiae,* there is competition for substrates between this yeast species and the lactic microflora, to the benefit of the yeast. Yeasts B2 and B3 present a more acidic metabolic profile (pH, TTA, lactic acid), with less competition between yeast and lactic acid bacteria, but also less yeast activity related to the yeast species present.

While the link between microbial flora and the characteristics of yeast is observable, the conditions of implantation of the microflora of a leaven and the nature of the interactions within the ecosystem between yeast and lactic acid bacteria are not explicit. Moreover, it should be noted that within a microbial species, there are variations that can be significant from one strain to another. This intra-species diversity makes it difficult to generalize from a yeast identity card, but it is also, within a yeast, a guarantee of resilience in the face of changes in environmental factors. It is the living character of a leaven that stands out here.

A more complete analysis, not addressed in this synthesis, integrating other biochemical compounds of the breads, such as volatile compounds, mineral elements, shows an effect of the factors leaven, terroir and variety (old/modern) on a number of markers (Thesis Elisa Michel).

Conclusion

The project has enabled the microbiological characterization of sourdough on the French territory of various bakers with a working environment free of industrial ferments and closely linked to peasant agricultural practices. Several new strains of yeast in sourdough were identified. The practices (artisan / peasant baker) have an impact on the biodiversity of sourdoughs. The diversity of practices and yeasts leads to bread characteristics, in terms of pH and acetic acidity, which theoretically do not allow the very restrictive regulatory framework for traditional French sourdough bread to be respected. It would be necessary for transparency towards the consumer to have a specific appellation such as "natural sourdough bread" that is representative of this diversity.

Secondly, and more specifically, the preparation of sourdough from different flours in different selected bakeries and its use in a reference bread-making process leads to various observations:

- **Sourdoughs** have shown **a greater effect than flours.**
- **The elaboration and conduct of the sourdough impacts the microbial flora and the activity of the chief sourdough.**

The number of refreshments must be sufficient and regular to allow sufficient and optimal activity of the sourdough, especially the yeasts, while maintaining the technological quality of the dough.

The environment of the bakery is confirmed to have a direct effect on the microflora that implants itself in a sourdough. The number of refreshments during the elaboration of the sourdough seems to be one of the determining factors of the quantitative and qualitative balance of the microbial ecosystem.

- **Flour varieties have different characteristics.**

The varieties chosen for the old and modern blends make it possible to clearly distinguish between two populations on the qualitative and quantitative aspects of protein and gluten, on the hardness of the grains, the rate of damaged starches and on the grain size of the flours and are consistent with the differences often observed between the two types of wheat. Our sampling therefore allows us to clearly characterize and distinguish between Modern and Ancient Blends representative of the diversity of wheat used by organic farmers and artisan bakers. In this context, it was interesting to see if yeast could specifically bring added value in bread-making.

Different behaviours were observed during baking, especially in terms of hydration and elasticity of the doughs, which are more important for modern varieties.

On the other hand, within the framework of the implementation of a reference process common to all flours, there are no observed effects of old and modern varieties on the growth or volume of breads. On the other hand, the alveolar structure appears more or less regular depending on the variety.

-   **Sourdough affects the characteristics of dough and bread.**

The characteristics of yeast influence those of dough and bread: elastic strength, dough resistance, gas production, growth and alveolar structure of the bread.

The metabolites resulting from the activity of the yeast, in particular organic acids, are dependent on the yeast. It follows that, with identical processes, the characteristics of the bread such as pH, acidity, Fermentation Quotient, but also volatile compounds, are impacted by the leaven.

These characteristics associated with the physical properties of the dough (volume, alveolar structure), have an impact on the sensory qualities of the final bread.

Under the conditions of the study and implementation of the yeast, the nature of the microflora present influences the characteristics of the breads. However, this observation should be taken into account by conducting the bread-making test in a differentiated and adapted manner according to the characteristics and functions of the sourdough.

The conduct of the yeast requires a particular vigilance to the regularity of the practices and attention to the observation of the behaviour of the yeast. If the diversity of wheat has an impact on the characteristics of the flours, the baking environment and practices seem to have more influence on the identity and properties of the sourdough.

- **The nature of the causal link between sourdough and the characteristics of dough and bread remains hypothetical.**

Variations in the elasticity and strength, growth, honeycombing and volume of the loaves seem to be related to several factors or hypotheses at the same time.

A higher microbial population of lactic acid bacteria and yeasts can impact the fermentative activity and alveolar structure at the level of the nucleation origins of the alveoli. Similarly, the properties of the gluten network in relation to the original wheat will impact the alveolar structure.

The variable number of refreshes can have an impact on the oxidation of the constituents and on the proportion of acetic acid. These two phenomena affect the state of cohesion or aggregation of the gluten. A low lactic acid/acetic acid ratio, i.e. a lower acetate content would be associated with a higher elasticity of the dough. Acetic acid here could also be a marker of the nature of the microflora and the duration of fermentation. This type of influence seems to be independent of the origin of the wheat (ancient or modern), but more dependent on the leaven and its conduct.

In the end, the participatory approach applied in the project, allowed both experimentation close to the field, an appropriation of the results by professionals and also cross approaches between technology and more fundamental science.

The experiment shows the impact of the diversity of practices and environments on the characteristics and properties of sourdough, this diversity is a guarantee of product differentiation and diversification. A better knowledge of the possible links between agricultural practices, product processing practices, product functionalities and characteristics and the sharing of this knowledge is also an element of the sustainability of the sector by enabling professionals to adapt their practices.

In a context of questioning about the maintenance of global biodiversity, it seems essential to us to also consider this biodiversity at the microbial level both because it is the basis of global diversity and because this diversity maintains ecosystems which, applied to food production, could be favourable to the diversity of fermented products, their properties and associated metabolites. The consumption of these products, resulting from practices that promote diversity, would therefore have an impact on both overall biodiversity and the quality of food through its variety.

Main bibliographical references :

- CNERNA, 1977. The bread. Recueil des Usages des Pains en France CNRS, Paris, 243-306.

- Gobetti M. and Gänzle M. (2013) Handbook on sourdough biotechnology. Springer Science & Business Media.

- ITCF (1972). Wheat quality control, practical guide. ITCF, Paris. 131-134.

- Lhomme E., Dousset X. Onno B. (2016). Sourdoughs for bread-making: Microbiota and Functionalities. Ed, Editions Universitaires Européenne.

- LhommeE., Urien C., Legrand J., Dousset X., Onno B., SicardD. (2016) : Sourdough microbial community dynamics: An analysis during French organic bread-making processes; Food Microbiology, Volume 53, 41-50.

- Michel E. (2018). Dispersion, selection and role of microbial species of sourdough in French low-input bakery. Thesis University of Montpellier SupAgro.

- Michel E., Montfort C., Deffrasnes M., Guezenec S., Lhomme E., Barret M., Dousset X., Sicard D., Onno B. (2016). Characterization of relative abundance of lactic acid bacteria species in French organic sourdough by cultural, qPCR and MiSeq high-throughput sequencing methods. International Journal of Food Microbiology 239, 35-43.

- Onno B., Roussel P. (1994) Technologie et microbiologie de la panification au levain, in Les bactéries lactiques, Tome II, Ed LORICA (38410 URIAGE), 310-312.

- Roussel P., Onno B. and Sicard D. (2020) La panification au levain naturel - Glossaire des savoirs, Ed. QUAE.

- Tripto 2014 Doc. Experimental bakery test sheet. https://www.triptoleme.org/recherche-et-publications.

- Tripto 2014 Doc. Glossary bakery experiment. https://www.triptoleme.org/recherche-et-publications.

- Tripto (2017-1). Grain hardness measurement. https://www.triptoleme.org/recherche-et-publications.

- Tripto (2017-2). Paysblé milling experiment. https://www.triptoleme.org/recherche-et-publications.

- Tripto (2017-3). Tripto 2017 Art. experimentation bakery PaysBlé. https://www.triptoleme.org/recherche-et-publications

- Urien C., Legrand J., Montalent P. , Casagerola S. and Sicard D. (2019). Fungal Species Diversity in French Bread Sourdoughs Made of Organic Wheat Flour; Front. Microbiol. ; https://doi.org/10.3389/fmicb.2019.00201

- Zheng J., Wittouck S., Salvetti E. et al., (2020). A taxonomic note on the genus Lactobacillus: Description of 23 novel genera, emended description of the genus Lactobacillus Beijerink 1901, and union of Lactobacillaceae and Leuconostocaceae. https://doi.org/10.1099/ijsem.0.004107 ; open access DOI: https://doi.org/10.7939/r3-egnz-m294

I want morebooks!

Buy your books fast and straightforward online - at one of world's fastest growing online book stores! Environmentally sound due to Print-on-Demand technologies.

Buy your books online at
www.morebooks.shop

Kaufen Sie Ihre Bücher schnell und unkompliziert online – auf einer der am schnellsten wachsenden Buchhandelsplattformen weltweit! Dank Print-On-Demand umwelt- und ressourcenschonend produzi ert.

Bücher schneller online kaufen
www.morebooks.shop

KS OmniScriptum Publishing
Brivibas gatve 197
LV-1039 Riga, Latvia
Telefax: +371 686 204 55

info@omniscriptum.com
www.omniscriptum.com

OMNIScriptum